Current Perspectives in Human Physiology

Selected Readings

1998 Edition

Edited by

Linda Vona-Davis
West Virginia University

and

Lauralee Sherwood
West Virginia University

Wadsworth Publishing Company
I T P® An International Thomson Publishing Company

Belmont, CA • Albany, NY • Bonn • Boston • Cincinnati • Detroit • Johannesburg • London • Madrid •
Melbourne • Mexico City • New York • Paris • Singapore • Tokyo • Toronto • Washington

Printed in the United States of America
1 2 3 4 5 6 7 8 9 10

For more information, contact Wadsworth Publishing Company, 10 Davis Drive, Belmont, CA
94002, or electronically at http://www.thomson.com/wadsworth.html

International Thomson Publishing Europe
Berkshire House 168-173
High Holborn
London, WC1V 7AA, England

International Thomson Editores
Campos Eliseos 385, Piso 7
Col. Polanco
11560 México D.F. México

Thomas Nelson Australia
102 Dodds Street
South Melbourne 3205
Victoria, Australia

International Thomson Publishing Asia
221 Henderson Road
#05-10 Henderson Building
Singapore 0315

Nelson Canada
1120 Birchmount Road
Scarborough, Ontario
Canada M1K 5G4

International Thomson Publishing Japan
Hirakawacho Kyowa Building, 3F
2-2-1 Hirakawacho
Chiyoda-ku, Tokyo 102, Japan

International Thomson Publishing GmbH
Königswinterer Strasse 418
53227 Bonn, Germany

International Thomson Publishing Southern Africa
Building 18, Constantia Park
240 Old Pretoria Road
Halfway House, 1685 South Africa

Production services: Dean W. DeChambeau
Cover Design: Seventeenth Street Studios
Cover Photo: Image of acetylcholine by Michael W. Davidson, National High Magnetic Field
Laboratory, Florida State University

ISBN 0-314-20640-X

Current Perspectives in Human Physiology

1998 Edition

Contents

Preface

Nowhere in the world of science is there more relevance to our everyday lives than in the study of physiology. Many of the concepts of physiology were developed from the fundamental principles of biology, chemistry, and physics. The diversity of physiology is reflected in the wide variety of research reports published in scientific journals. Thus, it becomes a lifelong challenge for teachers and their students to advance their understanding of physiology by reviewing the current literature.

This collection of newsworthy articles provides a more readable and informative approach for students introductory study of physiology. These forty-nine articles were carefully selected from general-interest periodicals and science magazines and organized into seven topics special interest to the student. Each article begins with a brief summary of the ideas presented in the selection and ends with a set of questions to help identify the key points of discussion.

Acknowledgments

The editors wish to express their sincere thanks to the many magazines and journals that allowed us to reprint their articles. ❏

Section One

Life in the Cell

1 *How can minute quantities of a single molecule elicit a dramatic response from an organism without ever entering its cells? This question is asked frequently by scientists interested in the process of signal transduction, a pathway from membrane to nucleus. It all begins when proteins on the cell surface bind with messenger proteins sent there specifically to initiate a cellular response. These proteins impart a cascade of connections only to end at the cell's nucleus, thus activating or deactivating the genes that reside there. Controlling any part of the signal pathway has remarkable potential as a tool for gene therapy. To many, tracking this event has become a lifelong adventure with stories that all seem to converge on a single idea about signal transduction.*

Conversations in a Cell

Gary Taubes

Discover, February 1996

Stuart Schreiber is discovering just how a cell talks with the outside world. The details of the process were long a mystery. But these days they're signaling the start of a revolution.

Twelve years ago, when Stuart Schreiber was 28, he earned the questionable distinction of being the first chemist to synthesize a naturally occurring compound call periplanon-b, the sex attractant of the American cockroach. Schreiber says he was drawn to periplanon-b because of its geometric beauty. Having synthesized the molecule in his Yale laboratory, though he decided he might as well pursue the obvious experiment, so he descended into the chemistry building basement.

"I went down with one of my graduate students, a flashlight, a shoe box—it was all pretty primitive—and I had a book showing pictures of different kind of cockroaches," he explains. "But I had trouble distinguishing the male from the female. I was read-

ing that the markings on the leg would distinguish them, but I could never do that. So I found some American cockroaches—the big ones—and I didn't know if they were male or female, but only the male responds to this periplanon-b. It was very exciting; you could take extraordinarily tiny quantities of the stuff, picograms, and puff them into the air, and these cockroaches would flap their wings and stand up. Only half of them would do it, and that's how you knew which were the males."

The news reports of Schreiber's synthesis were less than adulatory—*Esquire* magazine gave him a Dubious Achievement Award for creating a cockroach dating service. Still, Schreiber's experience with cockroach sexuality was the spiritual beginning of what may become a revolution in modern medicine and biology. It led Schreiber unwittingly from his mundane life as an organic chemist into the world of cell biology and the study of how signals are passed to living cells from the outside world. Little was known about that process, called signal transduction, when Schreiber sent his roaches into a tizzy. Researchers saw that chemicals floating in the body outside the cells somehow caused activity among the cells' genes; in the early 1980s, though, they had only just begun to ask why.

They knew that though DNA acts like an encyclopedic how-to manual for a cell, it is kept squirreled away in the nucleus, surrounded by material called cytoplasm, and sealed up tightly in the cell membrane. Tens of thousands of receptors stud the membrane, waiting for messages for the world outside the cell. The messages and receptors take the form of proteins, and their union is the first step in a pathway that acts like an electric circuit, carrying signals to the nucleus.

First a single molecule (periplanon-b, for instance) sidles up to a receptor on the external membrane of a cell. By contacting and binding to the receptor, it sets off a Rube Goldberg chain reaction of events inside the cell. There another molecule combines with a neighboring compound to trigger the next connection, and the signal is passed from chemical to chemical until it reaches the nucleus, activating the genetic mechanisms and the DNA inside. As a result, the cell might divide and replicate, or, if it's already busy dividing, it might stop. It might die or differentiate into some other kind of cell. It might pour out molecular messengers to activate more signal pathways in other cells.

But beyond knowing that chemicals had to be alerting one another within the cell, researchers had no idea what was going on, in detail. Not until 1993 were they able to describe even a single pathway completely, from membrane to nucleus. And that description was the work of researchers at a dozen labs.

Now Schreiber, along with Jerry Crabtree of Stanford and researchers at their two labs, thinks he has discovered a systematic way to reveal all the interactions that make up any pathway. Better still, the group's method may allow them to co-opt the signaling machinery inside cells, turning it on and off at will. The incipient technology holds promise as a new weapon for fighting a range of diseases, from cancer to cystic fibrosis and other developmental disorders, and Schreiber and Crabtree have

become two of the most widely read and cited researchers in all of science.

The story of their discovery is one of those medical mysteries in which the attempt to answer a single biological question leads to a discovery of revolutionary importance. It starts with Schreiber and his aesthetic obsession with molecules. Although in recent years he has been called "off-scale brilliant" and "clearly the leading chemist of his generation," Schreiber was far from a child prodigy. He spent his high school days working at a pizza parlor 60 hours a week and, by some accounts, partying the rest of the time. He went to college, at the University of Virginia, only because the sister talked him into it. His understanding of education was so limited that when he walked into a required freshman chemistry course and saw his classmates diligently taking notes in notebooks, he wondered how they knew to buiy notebooks and how they knew to take notes.

Then he experienced a revelation. "I discovered chemistry," he says. "It was my first learning experience: it was just beautiful. I was attracted to the aesthetics of chemistry, the shapes of molecules, the orbitals, the geometry. I latched onto organic chemistry, read everything I could get my hands on, took every graduate school course there was." He went on to get his Ph.D. at Harvard, where he studied with Nobel Prize-winning chemist Robert Burns Woodward, famous for synthesizing extraordinarily complex molecules. Woodward passed on his passion for that scientific art.

By the time he was 25, Schreiber was an assistant professor at Yale, where he did his cockroach work and started asking questions more relevant to biologists than chemists. At first they were about periplanon-b and male cockroaches: How could such tiny quantities of single molecule elicit such a dramatic response form an organism without ever entering its cells? Or as Schreiber puts it, "How on earth does that work?

• • •

Before Schreiber got an answer to this question, though, he was side-tracked further into hard-core biology. One of his Yale colleagues, Robert Handschumacher, identified the exact spot on human cell membranes where a compound known a cyclosporine would attach itself. Cyclosporine, which had been found in Norwegian dirt samples near the Arctic Circle, was considered extraordinarily useful for controlling the immune system.. Not only could it help prevent the body from rejecting transplanted organs, but it seemed to work miracles on autoimmune diseases—diseases in which the immune system attacks the body's own cells. But because cyclosporine often required toxic doses to function, researcher were on the lookout for similar but non-toxic agents.

Handschumacher found the cyclosporine receptor—the one cellular protein to which cyclosporine would bind—by fishing with cyclosporine in a soup of proteins made from the thymus gland and seeing what stuck. The process was like taking one piece of a 1,000-piece jigsaw puzzle and dragging it through the other 999 pieces until one of them hooked on. He named the receptor cyclophilin, after its affinity for cyclosporine. When Handschumacher told Schreiber about the two chemicals, Schreiber found himself with a new obsession. He realized for the first time that the organic molecules he studied as a chemist were not so different from the proteins, like cyclophilin, that form much of the working components of human cells. "Proteins," he says, " are long strings of amino acids, but with side chains that can rotate and adopt different structures. As an organic chemist, someone who's used to studying the shapes of things, I thought I might really understand how these proteins take on shapes, and therefore how they can interact with one another."

Schreiber started a cyclosporine research program in his laboratory but was interrupted in 1986 by the announcement from a Japanese pharmaceutical company of a new chemical compound called FK506, that also appeared to be a powerful immune-system suppresser. Schreiber liked the idea of studying a new drug without cyclosporine's long history. With his students and colleagues, he quickly synthesized FK506 and then went fishing, as Handschumacher had, for whatever protein it might be binding to in the cells. When the found such a protein, they named it FKBP, for FK506 binding protein. Thought they didn't know FKBP's natural function in the body, it could clearly interfere with the immune response, at least when FK506 was around. But how exactly were the pair accomplishing that feat?

With this question, Schreiber took a giant step into cell biology—and into signal transduction in particular. Early studies of cyclosporine and FK506 suggested that both proteins somehow managed to interfere with an important signal pathway called the T cell receptor pathway. T cells are the primary defense mechanisms of the immune system. When chemicals known as antigens, found on the surfaces of foreign invaders, bind to the molecular receptors on the membrane of a T cell, the T cell receptor pathway is turned on, signaling the DNA in the cell nucleus to create and release a protein called interleukin-2. Interleukin-2 stimulates T cells all over the body to replicate; in essence it's an alarm call, telling the immune system to prepare for danger. With cyclosporine or FK506 in the neighborhood, though, T cell replication ceases, antigens or no antigens.

When Schreiber began probing it, the T cell receptor pathway was an unknown. But Schreiber wanted to understand how FK506 and cyclosporine shorted out this pathway, and to do that he and his colleagues had to trace its wiring from beginning to end. "You literally do it by looking at one molecular interaction after another," Schreiber says. "You establish one, and then you pull out the thing it interacts with and ask what that one binds to. It is actually that simple. The complexity comes in when you realize that there are a lot of things binding to a lot of other things simultaneously, and you have to dissect the system to get one interaction at a time."

Schreiber was not alone in studying the T cell pathway. At Stanford, Jerry Crabtree was approaching the same problem from a different point of view. Crabtree, who grew up on a West Virginia farm, studied medicine in Philadelphia before moving on to research. He learned the nuances of cloning and DNA technology at the National Institutes of Health during the 1970s—at the time, says Crabtree, "about as exciting a place as anywhere could possible be."

Crabtree's interest was in how genes are turned on and off during in organism's development. By the end of the 1980s he was studying T cell activation. Simply put, he says, when T cells are activated by an antigen, for the next two weeks they will proceed step by step "through a very precise preprogrammed sequence of events." Some 200 to 500 genes will activate, one after another, ticking off like soldiers standing up to be counted, and will do so the exact same way each time, each gene producing its own particular protein. To make all these proteins, the immune system needs about two weeks—the same amount of time it takes the body to handle most infections. That's why serious infections such as pneumonia seem to last just that long. "The question was," says Crabtree, "How are those genes turned on at the right time?"

Crabtree and his researchers had two pieces of information to work with. Once a cell has passed a critical point in the activation process, which happened within an hour, it was committed to it. Take away the stimulus—the antigen—and the cell would still tick along through its two-week sequence. Furthermore, other researchers had established that T cells were sensitive to cyclosporine and FK506. Add either to the cell, and the activation process never starts.

Because the production of interleukin-2 (Il-2) is an important first step in this sequence, Crabtree started his research there. "The activation of that IL-2 gene," he explains, "would commit a cell to this process. That was the molecular core of the commitment. So the question boils down to: What turns on the IL-2 gene?" He discovered that a molecule known as nuclear factor of activated T leucocytes (NF-AT) made the DNA in the nucleus produce IL-2. NF-AT was made from two components, one in the cytoplasm outside the nucleus, the other in the nucleus. When an antigen attached itself to a T cell receptor, something in the signal pathway sent the component in the cytoplasm, NF-ATC, swooping down into the nucleus to join up with its counterpart, NF-ATN, and turn the gene on. Moreover, it seemed to be this particular liaison—NF-ATC with NF-ATN—that cyclosporine and FK506 shut down.

• • •

Crabtree published his findings in 1991. By then Schreiber had moved to Harvard; after reading the

paper, he called Crabtree, and the two agreed to collaborate. They set out to discover exactly what FK506 and cyclosporine were doing inside the cell. Schreiber speculated that it may not have been FK506 alone that created the therapeutic effect, or even FKBP, but somehow the two together. He knew from his studies that when FK506 bound to FKBP, the pair created a nice bumpy surface that looked as if it would bind neatly to some other molecule with complimentary indentations. That surface, says Crabtree, "could be exerting the therapeutic effect."

But if so, what was sticking to the surface? Schreiber and Crabtree got the hint they needed from Jeff Friedman, a graduate student working down the hall from Crabtree on a different project involving cyclosporine. Friedman found that when cyclosporine and cyclophilin combined, they bound themselves to yet another protein. Then, when one of Schreiber's postdocs at Harvard, Jun Liu, showed that the new protein was something called calcineurin, they knew they'd hit the jackpot, because calcineurin also bound to the combination of FK506 and FKBP.

Calcineurin was a key molecule in the T cell pathway. When it performed properly, it was responsible for sending NF-AT to work on the DNA. Take calcineurin out of the circuit—the combination of either FK506 and FKBP, or cyclosporine and cyclophilin did the trick nicely—and the wiring diagram was missing a crucial connection. In the T cells, FK506 allows two proteins that normally have nothing to do with each other—calcineurin and FKBP—to bind together, so cyclosporine and FK506, says Schreiber, act like a kind of molecular glue.

Natural versions of those gluelike molecules are vital for signal pathways to function. While the proteins that make up these signal pathways are always present in the cell, they will not normally interact until signaled to do so. That signal can take one of two forms: proteins, for instance, can change the shape, an event initiated by other proteins adding or snatching away a few small charged molecules. The new configurations allow the proteins to dock in the docking sites of others and pass their signals along. At one time, biologists thought most cellular interactions proceeded this way.

But researchers have now demonstrated that proteins can also pass along molecular signals just by getting close enough to latch onto each other ("dimerizing," in the lingo of biologists), or by latching onto two other molecules—receptors, for instance—and pulling them close together. This is the proximity effect, and it's how FK506 does its magic.

Now says Schreiber, researchers realize how prevalent the proximity effect is within the cell. Because signal pathways employ this mechanism so extensively, he says, it struck him, as it struck Crabtree, that if they could create molecules to emulate FK506 and work as molecular glue, they would "gain control over the pathways."

Crabtree says, "We thought maybe we could control the way proteins work by making some small organic molecule and just physically proximating two proteins." They would build a molecule called dimerizer, shaped like a little barbell or a two-headed key. One end would latch onto one protein target and the other to another, bringing the two proteins close enough to do whatever they would do naturally.

• • •

It took Schreiber's and Crabtree's labs some 18 months to pull it off; they tried different drugs and compounds, putting molecules together every which way. Finally they concocted a dumbbell-shaped molecule by sticking two FK506s together. The chemical which they call FK1012, was so similar to the components of the cell membrane that it would melt into the membrane and emerge on the other side. Once inside the cell, it would latch onto two proteins "and tie the things together," says Crabtree. "The physical proximity of that signaling unit would lead to some biological response."

This simple approach offered the researchers what looked like limitless power. Now, say Crabtree, "there's almost no level of biological control within the cell that we can't approach, beginning at the cell membrane and going all the way down to the nucleus." They have used their molecular dimerizers

to turn on the signal pathways below the membrane rather than relying on hormones to do it outside the cell. They've turned on signal pathways from the middle of the wiring diagram and even headed straight to the DNA in the cell nucleus and activated the genes directly, all of which used variations on their tiny dumbbell to induce a proximity effect. The variations are found by trial and error, but Schreiber's lab has developed several mechanisms to accelerated the hit-and-miss process, allowing the researchers to come up with dimerizers that work wherever they choose on the signal pathways.

Controlling signal pathways has remarkable potential. Schreiber and Crabtree say they can use their technology to activate or deactivate genes at will, by switching on or off the pathways that control them. They can also pull proteins out of pathways at any time, to see how that affects development. The technique is simple. They create molecules that act like FK1012, binding to the protein in which they're interested; these molecules, in Schreiber's words, "cause it to no longer function, and gum up its activity site." Then the researchers can learn the protein's purpose by triggering cell development or replication and seeing what the cell does without it.

They're learning to create molecules that will drag a protein into a proteus zone, the cell's internal garbage incinerator, in which it is rapidly destroyed. If this works in animals as well as it does in the laboratory, it should allow the researchers to study what a protein does on a particular day, or even in a particular hour, of development. "We could say what's happening at, say, day 14 after conception for a particular mouse protein," explains Crabtree, "by adding the dimerizer, making the protein disappear for a short time, and determining the consequences for development."

Eventually, Schreiber and Crabtree hope to be able to use their methods on human cells. The human body is prey to a host of diseases caused by missing or defective genes—sickle-cell anemia, for instance, and cystic fibrosis. Genetic engineers would like to insert genes into cells or replace defective genes with new ones. There are also infectious diseases that can be fought by prompting cells to create proteins, such as interferon, that stimulate the immune system.

While genetic engineers have been learning how to insert new genes into human cells and then get them back into into the body, they have yet to conquer the problem of turning the genes on when they're needed, or turning off genes that are producing harmful proteins. With their specifically engineered dimerizers, Crabtree and Schreiber may have found a way to do just that: FK1012 will permeate cell membranes and can be designed to turn on or off the pertinent signal pathways, and do it in dosages that can sit quite comfortable in a tiny pill. "If you give a tiny, tiny bit of FK1012," says Crabtree, "much less than you would give in aspirin, for example, it will go all over the body and get to the right places."

How far their revolution will go is anyone's guess. Schreiber and Crabtree have licensed the technology to a Boston-based biotech company, Ariad, that's working on everything from turning on proteins that will dissolve blood clots to gaining control over signal pathways that lead to growth inhibition or programmed cell death. The reason for the latter is cancer. "Using gene therapy to allow an organic molecule to kill only tumor cells would obviously be very significant," say Schreiber. "The new challenge is to get DNA into tumor cells, and then target those genes to kill the cells. If desired you can solve that problem once, then the organic molecule will work continuously."

Schreiber and Crabtree admit that their approach entails a host of challenges. But with the new approaches that open up as cell biology converges—in great part through their work—with fields like structural biology and Schreiber's own chemistry, the vision of overcoming those difficulties becomes ever clearer and sharper. ❏

Questions

1. What happens in signal transduction?

2. What are proteins?

3. How could controlling signal pathways have therapeutic value?

Answers are at the back of the book.

2   *Organs such as the cortical collecting duct of the kidney, the distal colon and the lungs all possess a unique epithelial sodium channel. Named the amiloride-sensitive epithelial sodium channel (ENaC), it influences a variety of different physiologic functions in the body from fluid volume regulation to mechanotransduction. So extensive is this channel's reach that it may play a significant role in the outcome of salt-sensitive hypertension and acute respiratory distress syndrome. Using molecular biology techniques to produce mutations of ENaC found in the lung, scientists are working to explain its role in the pathophysiology of cystic fibrosis.*

Epithelial Sodium Channels: Their Role in Disease

Bernard Rossier

News in Physiologial Science, April 1996

The amiloride-sensitive epithelial sodium channel (ENaC) is the major rate-limiting step for electrogenic sodium absorption in several major organs: the distal part of the renal tubule (mainly the cortical collecting duct), the distal colon, and the lungs. The primary structure of this highly selective, low-conductance channel was defined in 1994;[1] it is a heteromultimeric protein composed of three homologous subunit (α, β and γ). The channel is a member of a novel gene superfamily that, by influencing ionic conductances, participates in the regulation of a number of major variables and functions, such as cellular and extracellular volume, blood pressure, fluid volume in the lungs, neurotransmission, and taste transduction and mechano-transduction.

Until recently, the pathophysiological implications of ENaC were not fully appreciated because specific probes had not been developed to detect mutations in ENaC or to measure gene expression in various organs and tissues. Now, however, just 2 years after elucidation of the structure of ENaC, it appears that the clinical implications of this channel are wider than had been anticipated. Two major disorders in which mutations of ENaC are involved are Liddle's syndrome in the humans[3] and acute respiratory distress syndrome (ARDS) in a knockout mouse model.[2]

Liddle's syndrome (also called pseudohyperaldosteronism) is a monogenic, salt-sensitive form of hypertension that is inherited as an autosomal dominant trait. Careful genetic investigations in human patients have thus far revealed several mutations that lead to this syndrome;[3] truncation or missense at the carboxy-terminus of the β-subunit (Lifton's group and, more recently, Sasaki and his colleagues) and at the carboxy-terminus of the γ-subunit (Lifton's group). When expressed in the *Xenopus* oocyte system, these mutated genes lead to hyperactive ENaCs, an observation that fits well with the idea that salt-

Reprinted with permission from *News in Physiological Sciences*, Vol. 11, April 1996, p. 102.

sensitive hypertension is caused by inappropriately high reabsorption of sodium in the distal part of the nephron. Conversely, Lifton et al. have also described mutations of ENaC in several human pedigrees suffering from a salt-wasting syndrome (type 1 pseudohypoaldosteronism). Expression of these mutant genes in the *Xenopus* system leads to hypoactive ENaCs, although not to total loss of channel activity.

Complete loss of ENaC, however, has now been engineered in knockout mice by inactivating the γ-subunits.[2] the results are striking: amiloride-sensitive electrogenic sodium transport was abolished in airway epithelia cultured from these knockout mice (—/—), and neonates of this mouse strain developed ARDS, with failure to clear fluid from their lungs leading to death within 40 h after birth. This finding emphasizes the critical role of ENaC in the adaptation of the lungs of newborns to air breathing by clearing fluid from the lungs (which were filled with fluid up to the moment of parturition). Recently, Stutts and colleagues have emphasized the role of interactions between cystic fibrosis transmembrane conductance regulator (CFTR) and ENaC in the physiopathology of cystic fibrosis.

In conclusion, abnormalities of ENaC have become an important field of physiological and clinical investigation. The diseases here described, pseudohyper- and pseudohypoaldosteronism and ARDS, are prime examples of where clinical abnormalities have stimulated physiological investigation, and vice versa. No doubt other diseases will soon be ascribed to mutations in the ENaC gene, and each newly described abnormality will greatly help the physiologist to identify and characterize the most important functional domains of this important sodium channel.

References
1. Canessa, C. M., L. Schild, G. Buell, B. Thorens, I. Gautschi, J.-D. Horisberger, and B. C. Rossier. Amiloride-sensitive epithelial Na$^+$ channel is made of three homologous subunits. *Nature Lond.* 367: 463-467, 1994.
2. Hummler, E., P. Barker, J. Gatzy, F. Beermann, C. Verdumo, R. Boucher, and B.C. Rossier. Early death due to defective neonatal lung liquid clearance in α-ENaC-deficient mice. *Nature Genet.* In press.
3. Shimkets, R. A., D. G. Warnock, C. M. Bositis, C. Nelsonwilliams, J.H. Hanson, M. Schambelan, J.R. Gill, S. Ulick, R.V. Milora, J. W. Findling, C. M.Canessa, B.C. Rossier, and R.P. Lifton. Liddle's syndrome; heritable human hypertension caused by mutations in the ß-subunit of the epithelial sodium channel. *Cell* 79: 407-414, 1994. ❏

Questions

1. What is the amiloride-sensitive epithelial sodium channel (EnaC)?

2. Where are these channels found?

3. What is Liddle's syndrome?

Answers are at the back of the book.

3   *Any athlete will agree that muscles are the contraction specialists of the body. Excitation of skeletal-muscle fiber by its motor neuron brings about contraction through a series of molecular events. Calcium ions signal the protein in muscle cells called troponin-C to change its shape. But how does this occur? By examining the protein up closely using nuclear magnetic resonance spectroscopy, scientists discovered that the positively charged calcium acts as the lure to attract a negatively charged amino acid glutamate in closer to it causing the protein to change shape. Such knowledge could aid in the design of drugs to strengthen muscle contractions of undamaged heart muscle after a heart attack.*

Flexing Muscle with Just One Amino Acid

Robert F. Service

Science, January 5, 1996

The workings of muscles interest more than just athletes: For years, scientists have been tracing the cascade of molecular events that trigger muscle contraction. One key player in the sequence is a protein in muscle cells known as troponin-C, which responds to a chemical signal—the release of calcium ions form within the cell—by changing its shape. The contortion alters its chemical interactions with neighboring proteins, and these interactions eventually lead to cell contraction. But just why troponin-C undergoes this crucial shape change has remained murky.

Two weeks ago, however, a group of Canadian researchers, flexing some investigatory muscle of their own at 1995 International Chemical Congress of Pacific Basin Societies in Honolulu, Hawaii, may have cleared up this mystery. The researchers, led by Brian Sykes, a biochemist at the University of Edmonton in Alberta, Canada, unveiled new studies showing that a single amino acid—glutamic acid, the 41st amino acid in the protein chain—controls this shape change by dragging a section of the protein toward a newly bound calcium ion; mutant proteins without the amino acid didn't budge. Troponin-C comes in two forms, one in skeletal and one in cardiac muscle. These studies were done on the skeletal form, but researchers believe the information may aid the design of drugs, known as calcium-sensitizing drugs, intended to treat heart attack victims by strengthening the contractions of undamaged heart muscle cells.

"It's a very nice study," say Walter Chazin, a molecular biologist at the Scripps Research Institute in La Jolla, California. The new result is the first to suggest that a large shape change can be controlled by the identity of just a single amino acid, he says. But although Chazin finds the evidence compelling, he cautions that some uncertainty remains, because the mutation could also be altering the protein's response by affecting the way it binds calcium.

Sykes's groups has been pursuing the relationship between troponin-C's structure and function for some time. Their new result comes just 3 minutes after they completed the first description of the protein in its calcium-bound state. Past x-ray crystallography studies of the unbound protein had shown that the amino acids in the key regulatory section of the molecule are woven into a series of the five connected helixes and a pair of loops. Researchers suspected that he calcium ions were bound inside the loops and played a role in the protein's shape change. But scientists were unable to confirm these suspicions, because suspicions, because they couldn't crystallize the protein in it calcium-bound configuration to study it with x-ray.

But in the September issue of *Nature Structural Biology*, Sykes and his colleagues Stephane Gagné, Sakae Tsuda, Monica Li, and Larry Smillie addressed the question with a different structure-determining technique, known as nuclear magnetic resonance (NMR) spectroscopy, that doesn't require crystallization. The NMR technique, which determines the position of atoms within a protein from the way they resonate in a magnetic field, confirmed that calcium binding takes place inside the two loops.

The bound structure also revealed a clue to how this might trigger the shape change. The calcium-bound structure showed that a calcium ion in one loop was sitting near a glutamic acid in a neighboring helix; in the unbound structure, the two are farther apart. That difference suggests that upon binding to the protein, the positively charged calcium attracts the negatively charged glutamic acid, which pulls the helix along with it, forcing the protein to change its shape, says Sykes.

But this circumstantial evidence didn't eliminate the possibility that other amino acids were involved in the shape change as well. So in their latest study, the Alberta researchers created a mutant version of the protein in which the key glutamic acid was changed to an alanine, another amino acid—but one with a neutral charge. When the team studied the structure of the mutant protein, they found that it no longer altered it shape even after the two calcium ions bound inside their loops. With no electronic attraction, Sykes suggests, there is nothing forcing the helix to change its position.

The structural information may provide clues to shape changes in the cardiac form of the protein, and thus help to design drugs that strengthen heart muscle contractions by keeping that form in its calcium-bound position, says R. John Salero, a physiologist studying such drugs at the University of Illinois, Chicago. But structural studies on the cardiac form have yet to be completed.

Chazin also noted one complication in studies of the skeletal form. The glutamic acid is normally a key link in the protein's framework that holds the calcium in place, and the mutational change may change the way calcium sits in the binding pocket. That in turn could prevent the shape change by altering the way in which the calcium interacts with neighboring atoms.

But so far Sykes sees no sign of that. "I would say our data show the binding pocket is still as it was," say Sykes. "And there isn't any evidence that the calcuim is binding anywhere else." And Chazin says that seeing the full NMR data of the new protein once it's published should help settle the matter. "If he's got the right answer," says Chazin, "then he's on to something very big." ❏

Questions

1. What does troponin-C do in skeletal muscle?

2. Which amino acid controls the shape of the troponin-C?

3. What does nuclear magnetic resonance (NMR) spectroscopy determine?

Answers are at the back of the book.

4   *Molecular geneticists and public health officials ar combining forces to elucidate a highly common but underrecognized disorder called hemochromatosis. Affecting approximately one million Americans, hemochromatosis is a condition of abnormal iron absorption across the gut epithelium. Because the body's ability to excrete iron is limited, iron overload develops which can lead to organ injury and disease. unfortunately, most cases of hemochromatosis go undiagnosed even though an effective and inexpensive treatment exists. Through education and screening, the Center for Disease Control and prevention hopes to identify those persons at high risk as the first step toward decreasing the morbidity and mortality associated with this genetic malady.*

Hemochromatosis: The Genetic Disorder of the Twenty-first Century

James C. Barton and Luigi F. Bertoli

Nature Medicine, April 1996

Recent molecular genetics and public health efforts are aimed at understanding, detecting and treating this common but underrecognized disorder.

Hemochromatosis, a condition of abnormal iron absorption, is among the most common genetic maladies in humans, far exceeding the combined incidence of cystic fibrosis, phenylketonuria, and muscular dystrophy. Described between 1877 and 1889 in a few autopsy cases in France and Germany, hemochromatosis has largely gone unheeded for most of the twentieth century mainly because of the mistaken notion that it is a rare clinical disorder. However, there are 1 in 200 homozygous and 1 in 8 heterozygous individuals carrying the hemochromatosis gene(s) among the Caucasian population.[1] Currently, there is no formal recom- mendation from major medical bodies concerning hemochromatosis, nor has a national research effort been undertaken to identify the responsible gene(s). A few weeks ago a panel of scientific experts met in Atlanta to assist the Centers for Disease Control and Prevention (CDC) in developing national guidelines and programs to detect, to prevent and to treat iron overload diseases. For the first time, hemochromatosis was on the agenda, and meeting participants made several recommendations appli- cable to public health and clinical practice as a first step toward decreasing morbidity and mortality as- sociated with this disease. The CDC is now finaliz- ing guidelines for the screening of persons at high risk with the goal of ultimately screening all adults.

Iron is normally absorbed across the gut epithe- lium, primarily in the duodenum and upper je- junum, and is regulated by total-body iron stores, in a manner that is still not well understood. In the average diet, 15-20 mg of iron is available, yet each day only approximately 1 mg of iron is absorbed by males and 2 mg by females of reproductive age. In hemochromatosis, iron absorption by gut epithelial cells is inappropriately high. The diagnostic hallmark of homozygosity is persistent elevation of the iron saturation of serum transferrin to ≥60% in males and ≥50% in females.[2] Because mechanisms to excrete iron are limited, iron overload develops in most homozygotes. Among the 13% of American Caucasians who are hemochromatosis heterozygotes,[3] some will develop clinically apparent iron overload disease, usually when iron or alcohol intake is increased, or when alleles for hereditary anemias or porphyria cutanea tarda (which also increase iron absorption) are independently inherited.[2,3] Progressive iron deposition injures the liver, joints, pancreatic islets, other endocrine organs and the heart. Weakness, fatigue, hepatic cirrhosis, primary liver cancer, arthropathy, diabetes mellitus, hypogonadotrophic hypogonadism and cardiomyopathy caused by iron overload are often aggravated by other inherited, acquired or environmental factors.[2] Because intestinal absorptive pathway(s) do not function exclusively for iron, the absorption of cobalt, lead, manganese, and zinc are also increased in hemochromatosis, although the clinical significance of this phenomenon is still unclear.[4]

The molecular basis of aberrant iron absorption in hemochromatosis remains unknown, and no one has yet isolated the gene responsible. However, studies have shown that the gene for hemochromatosis is located on chromosome 6 and maps to the human leukocyte antigen (HLA) region, analogous to the major histocompatibility complex (MHC) in the mouse.[2] Rothenberg and colleagues have recently identified regulatory motifs (βGAP) homologous to the β-globin promoter that seem to control the expression of class I genes located within the murine MHC. As the hemochromatosis gene maps to a similar region on chromosome 6 in humans and class I genes associated with the βGAP sequence are expressed in the gastrointestinal tract of the mouse (the primary site of iron absorption) it seemed possible that these genes might be involved in iron

absorption. β$_2$-microglobulin knockout mice, which have altered expression of class I gene products, develop an iron overload condition similar to hemochromatosis. In a recent issue of the *Proceedings of the Nation Academy of Sciences*, Rothenberg et al.[5] present data showing that β$_2$-microglobulin knockout mice absorb and store an abnormal amount of iron in various organs compared with contol animals. These results indicate that β$_2$-microglobulin-associated proteins and perhaps unique class I genes themselves are involved in the control of intestinal iron absorption and possibly in hemochromatosis.

The frequency of the hemochromatosis gene among Caucasians implies that heterozygosity confers an evolutionary benefit. To absorb increased fractions of ingested iron may increase chances of survival during intervals of food deprivation, or may permit child-bearing females to impart greater quantities of this vital metal to their offspring. In addition to providing a nutritional advantage, the hemochromatosis gene may also increase resistance to a variety of infections.[6] African iron overload, a genetic disorder that also causes increased iron absorption, affects more than 20% of sub-Saharan native peoples[7] and other persons of African origin.[8] In contrast to hemochromatosis, it is not HLA-linked and may have distinct clinical and pathological features, suggesting that additional hereditary disorders which augment iron absorption[9] will be discovered when systematic evaluations of iron status in large population groups are performed.

Hemochromatosis homozygosity affects approximately one million Americans and is the most common known genetic disorder among Caucasians. However, most cases remain undiagnosed in routine clinical practice because of the misconception that hemochromatosis is rare, a misunderstanding of the diagnostic criterion, lack of familiarity with the multisystem disease caused by iron overload, or the belief that only those persons with "bronze diabetes with cirrhosis" have hemochromatosis. The keys to early diagnosis and treatment are education of health-care providers and population screening.[10] Appropriately viewed as public health problems, iron overload disorders including hemochromatosis are common, well-defined, and have costly medical consequences. Screening and diagnostic tests are available and affordable, and have favorable performance characteristics.[2,10] The primary treatment for hemochromatosis homozygotes and iron-overloaded heterozygotes is phlebotomy, a relatively inexpensive and simple technique that involves removing blood weekly from patients until their iron stores are exhausted. Normal iron stores are maintained thereafter by periodic bleeding. With early diagnosis and treatment, all known complications of iron overload due to hemochromatosis can be prevented. Treatment is cost-effective and low-risk, and may benefit others (by providing blood products for transfusion)[2]

The formulation of strategies to solve many of the clinical, scientific and public health issues relating to hemochromatosis should ensure that this disease rightly takes its place as the genetic disorder of the twenty-first century.

References

1. Edwards, C.Q. *et al.* Prevalence of hemochromatosis among 11,065 presumably healthy blood donors. *N. Engl. J. Med.* **318**, 1355-1362 (1988).
2. Witte, D.L. *et al.* Practice parameter for hereditary hemochromatosis. *Clin. Chim. Acta* (in press).
3. McLaren, C.E. *et al.* Prevalence of heterozygotes for hemochromatosis in the white population of the United States. *Blood* **86**, 2021-2027 (1995).
4. Barton, J.C. *et al.* Blood lead concentrations in hereditary hemochromatosis. *J. Lab. Clin. Med.* **124**, 193-198 (1994).
5. Rothenberg, B.E. & Voland, J.R. β_2 knockout mice develop parenchymal iron overload: A putative role for class I genes of the major histocompatibility complex in iron metabolism. *Proc. Natl. Acad. Sci. USA* **93**, 1529-1534 (1996).
6. Brock, J.H. Iron in infection, immunity, inflammation, and neoplasia. In *Iron Metabolism in Health and Disease*. (eds. Brock J.H., Halliday, J.W., Pippard, M.J. & Powell, L.W.) 353-390 (Saunders, London, 1994).
7. Gordeuk V. et al. Iron overload in Africa: Interaction between a gene and dietary iron content. *N. Engl. J. Med.* **326**, 95-100 (1992).
8. Barton, J.C., Edwards, C.Q., Bertoli, L.F., Shroyer, T.W. & Hudson, S.L. Iron overload in African Americans. *Am. J. Med.* **99**, 616-623 (1995).
9. Eason, E.J., Adams, P.C., Aston, C.E. & Searle, J. Familial iron overload with possible autosomal dominant inheritance. *Aust. N.Z. J. Med.* **20**, 226-230 (1990).
10. Edwards, C.Q. & Kushner, J.P. Screening for hemochromatosis. *N. Engl. J. Med.* **328**, 1616-1620 (1993). ❏

Questions

1. What is hemochromatosis?

2. How much iron is available and absorbed through the average American diet?

3. What is the current treatment for hemochromatosis?

Answers are at the back of the book.

5 *In the United States alone, about 1 in 400 African-American infants is born with sickle cell disease. People with sickle cell anemia have an abnormal type of hemoglobin called hemoglobin S. This genetic error makes the hemoglobin molecules stick together causing the red blood cells to become hard and sickle-shaped. The curved shape causes them to get stuck and plug up small vessels, interrupting blood flow to major organs. A painful crisis is the hallmark of the disease and to date, no FDA-approved treatment is available to prevent it from occurring. In the past, individuals with sickle cell anemia often died in childhood, however, antibiotics have lowered their risk for infection. Despite the absence of a completely effective treatment, life expectancy for individuals with sickle cell anemia has improved.*

New Hope for People with Sickle Cell Anemia

Eleanor Mayfield

FDA Consumer, May 1996

In tropical regions of the world where the parasite-borne disease malaria is prevalent, people with a single copy of a particular genetic mutation have a survival advantage. Over time, people from these regions have migrated, had children, and in some cases married each other. Some of their children inherit two copies of the mutation.

While inheriting one copy of the mutation confers a benefit, inheriting two copies is a tragedy. Children born with two copies of the genetic mutation have sickle cell anemia, a painful disease that affects the blood cells and is curable only in rear instances.

A recent trial sponsored by the National Heart, Lung, and Blood Institute (NHLBI) found that the drug Hydrea (hydroxyurea) significantly reduced painful episodes in adults with a severe form of sickle cell anemia. NHLBI is a component of the National Institutes of Health (NIH).

Hydrea is approved by the Food and Drug Administration to treat certain types of cancer. On the basis of the new clinical trial findings, FDA is encouraging Hydrea's manufacturer, Bristol-Myers Squibb, to apply for approval of the drug to treat sickle cell anemia.

"FDA will consider it a priority application so that sickle cell anemia can be added to the indications for Hydrea, if the data show that [the drug] is indeed safe and effective," said Anthony Murgo, M.D., a medical officer in the division of oncology at FDA's Center for Drug Evaluation and Research.

Genetic Defect Changes Cell Shape

The genetic defect that causes sickle cell anemia affects hemoglobin, a component of red blood cells. Hemoglobin's job is to carry oxygen to all the cells and tissues of the body. Red blood cells that contain normal hemoglobin are soft and round. Their soft texture enables them to squeeze through the body's small blood vessels.

People with sickle cell anemia, however, have a type of abnormal hemoglobin called hemoglobin S. (Normal hemoglobin is called hemoglobin A.) A genetic error makes the hemoglobin molecules stick together in long, rigid rods cause after they release oxygen. These rods cause the red blood cells to become hard and sickle-shaped, unable to squeeze through tiny blood vessels. The misshapen cells can get stuck in the small blood vessels, causing a blockage that deprives the body's cells and tissues of blood and oxygen.

When this happens "it's like having mini heart attacks throughout the entire body," said Duane R. Bonds, M.D., leader of the sickle cell disease scientific research group at NHLBI in Bethesda Md.

"A heart attack is painful because the blood flow to the heart is interrupted. In the sickle cell anemia, the blood flow can be interrupted to any of the major organs, causing severe pain and organ damage at the site of the blood flow blockage."

These painful "crises," as they are called, damage the lungs, kidneys, liver, bones, and other organs and tissues. Recurrence of these episodes is the most disabling feature of sickle cell anemia. They can cause leg ulcers, blindness, and many other health problems, depending upon where in the body the blood flow blockage occurs. A blockage in the brain can cause a stroke, which may result in paralysis or death.

The body, recognizing that the sickled cells are abnormal, destroys them at a faster rate than it can replace them. This causes a type of anemia, a shortage of red blood cells. Symptoms of anemia include extreme fatigue and susceptibility to infection.

Penicillin Treatment

An important breakthrough occurred in 1986 when an NHLBI-sponsored study found that young children with sickle cell anemia who took penicillin twice a day by mouth had much lower rates of *S. pneumoniae* infection than a similar group of children who received a placebo.

In 1987, an expert panel convened by NHLBI recommended that all infants born in the United States be screened for sickle cell anemia so that children with the disease could be identified early and offered treatment with penicillin. Forty-two states now have newborn screening programs, according to Bonds. NHLBI also recommends that affected infants get daily penicillin therapy beginning by the age of 3 months.

In the first year of life, children with sickle cell anemia are protected from blood flow blockages by the presence of fetal hemoglobin.

"Fetal hemoglobin is a hemoglobin that all of us produce before we're born," explained Bonds. Fetal hemoglobin physically blocks hemoglobin S. preventing it from forming the long, rigid rods that lead to sickling of the red blood cells. Several weeks before birth, however, the fetus's bone marrow usually begins to shut down the production of fetal hemoglobin and starts making adult hemoglobin instead. At birth, an infant's red blood cells contain roughly equal amounts of fetal and adult hemoglobin.

By the time the child is 6 months old, it has usually stopped making any fetal hemoglobin. As the level of fetal hemoglobin in the child's blood falls, explained Bonds, there is no longer anything to prevent the red blood cells from becoming sickle-shaped and getting stuck in the blood vessels, causing a painful crisis.

The symptoms of a painful crisis may be relieved by giving patients fluids and painkillers, said Lilia Talarico, M.D., a medical officer in the division of gastrointestinal and hematologic drug products in FDA's Center for Drug Evaluation and Research. However, no FDA-approved treatment is currently available to prevent painful crises from occurring.

Life Expectancy Improves

Despite the absence of an effective treatment, life expectancy for individuals with sickle cell anemia has improved, said Bonds, as a result of early identification through neonatal screening, early initiation of penicillin therapy, close medical monitoring, and early intervention to relieve the symptoms of a painful episode. A recent study found that half of all patients with sickle cell anemia survive into their 40s.

Rates of early death are highest among those with a severe form of the disease. "Between 10 and 15 percent of patients will have three or more painful crises per year," said Bonds. "The more crises you have per year, the greater your chances of dying prematurely. Your organs become more damaged when they are chronically not receiving enough blood and oxygen."

Some people continue to produce fetal hemoglobin throughout their lives. A study of the life expectancy of people with sickle cell anemia found that adults with high levels of fetal hemoglobin lived longer than those who had low levels.

People with sickle cell anemia who suffer strokes or infections, who are pregnant, or who must undergo surgery may be treated with blood transfusions. Risks of this treatment include the possibility of acquiring a viral illness such as hepatitis and the possibility of organ damage cause by iron overload.

Sickle cell anemia can be cured by a bone marrow transplant, which replaces the defective red blood cells with healthy cells from a donor. But a transplant is not a realistic option for most people with sickle cell anemia, according to Bonds, because of a shortage of compatible donors and because of the risks presented by the drug regimen that is required to prepare a patient for a transplant.

"First you give drugs to kill off the patient's marrow, then you do the transplant to replace the marrow." But the powerful drugs given to kill the patient's bone marrow can be dangerous for someone who has had a stroke or is at risk for stroke, she said.

In January 1995, NHLBI announced the successful conclusion of a five year, multicenter trial of Hydrea in the treatment of sickle cell anemia. The study involved 299 patients ages 18 and older who were recruited at 21 medical centers in the United States.

All patients had experienced at least three painful crisis in the year before they entered the trial. Half of the patients received Hydrea and half received a placebo. Neither the patients nor their doctors knew who was taking the drug and who was taking the placebo. Patients who took Hydrea had roughly half as may painful crises as those who took the inactive pill.

The trial had been scheduled to conclude in May 1995. However, scientists involved found the results so compelling that they stopped the study early and notified doctors of the results so that all patients who might benefit could be offered the treatment. A report of the trial's findings was published in the May 1995 *New England Journal of Medicine*.

Hydrea Studies

Hydrea is approved by FDA to treat certain types of leukemia and other cancers. Doctors have been interested in Hydrea for the treatment of sickle cell anemia for about 10 years, since pilot studies in humans showed that the drug could increase the level of fetal hemoglobin in red blood cells.

Because Hydrea is already on the market for other uses, it was unnecessary for FDA to issue a Treatment IND (investigational new drug) to make the drug available to patients with sickle cell anemia, said FDA's Murgo. A Treatment IND is a mechanism use by FDA to make investigational new drugs available to patients while they are under study.

As with other approved medications, doctors may prescribe Hydrea for sickle cell anemia if in their professional judgment a patient will benefit for the treatment.

"We are excited about the report [of the Hydrea clinical trial]," said Murgo. "But until the manufacturer [of the drug] submits the detailed data for us to review, [the agency] cannot approve the drug to treat sickle cell anemia."

Bonds, who was the project officer for the multicenter study, cautioned that Hydrea treatment is not appropriate for every patient.

"We only recommend it for patients over 18 who have had at least three painful crises in the previous year. The patients have to be monitored very carefully. The must have a blood test every two weeks to ensure that their blood count is not depressed to a level where they might be at risk for infection or bleeding."

Hydrea should not be prescribed for patients who are likely to become pregnant or who are unable to follow instructions regarding treatment, said Murgo.

Many questions about Hydrea in the treatment of sickle cell anemia remain unanswered, said Bonds. Doctors do not know what the most effective and least toxic dose of the drug is or whether taking it for many years presents health risks.

Some doctors prescribe Hydrea to treat polycythemia vera, a disease in which the number of red blood cells increases abnormally. Some evidence suggests that the drug may cause leukemia in a few of these patients, both Murgo and Bonds said.

However, they added, patients with polycythemia vera already have a higher-than-average risk of getting leukemia, so it is unclear whether the leukemia is caused by Hydrea or whether patients would have developed it anyway.

NHLBI is planning a five-year follow-up study of patients who took part in the Hydrea trial to see whether any of them develop leukemia or other problems that may result from long-term use of the drug. Another study, which began in January 1995, is looking at the safety and effectiveness of Hydrea in children ages 5 to 15 who have sickle cell anemia.

Research continues on other possible ways of reducing the occurrence of painful sickle cell episodes by increasing the production of fetal hemoglobin. For example, NIH scientists are studying whether a combined regimen of Hydrea and erythropoietin, a hormone that increases the production of red blood cells, is less toxic and more effective than hydrea alone. (Erythropoietin is licensed by FDA to treat anemia in certain patients.) Studies are also under way to determining the safety and efficacy of butyrate, and experimental drug that can stimulate production of fetal hemoglobin.

NHLBI recently funded three centers that will try to develop gene therapy for sickle cell anemia, said Bonds. "If you could replace the abnormal genes, you could cure the disease. However, there are significant technical problems involved in making gene therapy work."

Because of these problems, gene therapy is unlikely to be reality for many years, she said. Nevertheless, she said she is optimistic that new, effective treatments for sickle cell anemia will b developed in the future.

"I like to tell people that [the results of the Hydrea study] will one day be likened to [the discovery of] insulin or penicillin. Those drugs were the first major breakthroughs for the treatment of diabetes and severe bacterial infections, although other agents have since [been introduced] to treat those diseases." ❑

Questions

1. What characterizes sickle cell anemia?

2. How does sickle cell disease affect the flow of blood?

3. What causes the anemia associated with sickle-cell?

Answers are at the back of the book.

6 *Nerve cells need a constant supply of synaptic vesicles to package the neurotransmitters that are needed in signal transmission. To accomplish this feat, the plasma membrane forms a pit lined with the protein clathrin, beginning the process of endocytosis. Once the pit is completely formed into a sphere, the vesicle is severed from the membrane by a helical structure call dynamin, which wraps around the neck of the vesicle, pinching it off from the membrane. Scientists studying this unique protein have found this large GTPase to be necessary for the endocytosis of surface membranes in a variety of cell types, including neurons. Of particular interest is the role of dynamin in protein sorting and membrane fission using these helical structures.*

Ringing Necks with Dynamin

Regis B. Kelly

Nature, March 9, 1995

Synaptic vesicles in nerve cells are packages of neurotransmitters that are ejected each time a neuron fires. The stock of vesicles is then replenished rapidly by endocytosis of their membrane components from the plasma membrane and repackaged with neurotransmitters once inside the cell. An agent instrumental in sustaining the supply of vesicles, dynamin, has now been collared.[1,2]

Dynamin, a large GTPase, is needed for the endocytosis of surface membrane in a large number of cell types, including neurons.[3] Hinshaw and Schmid[2] show that it can adopt a helical structure in solution; Takei *et al.*[1] demonstrate that it is this helical structure that enables dynamin to wrap around the neck of membrane vesicles and help pinch them off from the plasma membrane. Dynamin- like molecules do not seem to be universally required for membrane fission. An ability to self-assemble into helical arrays, however, may be a common feature of dynamin and its homologues and could explain their physiological properties.

Endocytosis of membrane proteins from the plasma membrane involves three steps:[4] clustering of the membrane proteins by clathrin and its associated proteins, invagination of the membrane into a constricted clathrin-coated pit and, finally, severing of the neck to give a coated vesicle. *In vitro*, this sequence can be arrested at different stages, depending on which analogue of GTP is present.[5] The defining characteristic of the constricted-vesicle stage is that the neck allows small molecules to pass into the vesicle, but excludes protein. Structures that look like constricted vesicles are found at the surface of mammalian and *Drosophilia* cells in the presence of mutant forms of dynamin.[6-8] Although it seemed plausible that GTP hydrolysis by dynamin could be necessary for severing the neck of the coated pit, it was not clear just how it did it.

Dynamin is abundant in brain (1.5 per cent of total protein), and possesses, in addition to a GTPase domain, a pleckstrin-homology domain, and a carboxy-terminal proline-rich domain containing

Reprinted by permission from the author and *Nature,* Vol. 374, March 9, 1995, pp. 116–117.
Copyright 1995 Macmillan Magazines Ltd.

an SH3-binding motif.[3] The demonstration of its ability to self-assemble into long helical structures in the test tube[2] is complemented by the elegant electron micrographs from DeCamilli's Laboratory,[1] which show that dynamin surrounds the neck of the membrane. When permeabilised nerve terminals are incubated in the presence of the nonhydrolysable GTP analogue. GTP-γS, long invaginations are formed which are coated with spirals of dynamin. Clathrin-coated bulbs are often seen at the tips of the invaginations. The spiral of polymerised dynamin squeezes the membrane tube to an outside diameter of about 30 nm, and an internal diameter tight enough to prevent the passage of proteins. The GTP-associated form of dynamin is believed to ring the neck of the constricted pit:[1,2] GTP hydrolysis could change the configuration of the helix, squeezing the neck further until the membrane fission occurs in a sort of molecular garroting.

This pinching-off of membrane necks occurs frequently in biology, in events as diverse as cell division, budding of membrane-coated viruses, and vesicle budding from an intracellular organelle. If a molecule like dynamin is required in each case, then GTP hydrolysis should be necessary for membrane fission. This is not true for budding from the endoplasmic reticulum or Golgi, where GTP hydrolysis is required for vesicle uncoating and docking but not for vesicle formation. Dynamin may be able to help bud small vesicles from the plasma membrane but not from intracellular membranes because of the difficulty of curving a stiffer membrane. Plasma membrane differs fundamentally from intracellular membranes in that it is supported on the inside by cortical cytoskeleton and is attached to matrix and other cells on the outside.

Homologues of dynamin all have highly similar GTPase domains. One class of analogues, the Mx proteins, can confer resistance to viral infection. Until now it has been difficult to see a functional link between dynamin and homologues that regulate viral infection. An exciting clue comes from the observation that the mouse Mx protein can also polymerise into C-shaped and helical molecules, depending on the nucleotide concentration, that are 11 nm thick and 100–150 nm long, parameters similar to those found for dynamin-1. Thus the dynamin family may share a propensity to form filamentous aggregates, regulated like mictotubules by GTP hydrolysis. Perhaps Mx proteins interfere with virus infection by wrapping around cylindrical structures of about 25 mm diameter.

There is also an intriguing link between dynamin and VPS1 protein, a dynamin-like protein in yeast involved in protein sorting. Mutations in VPS1 send a vacuolar protease to the surface and a Gogli enzyme to the vacuole.[10,11] Most models of protein sorting require a mechanism for removing membrane proteins from an organelle, leaving behind the soluble protein content. Such segregation is needed, for example, to return low-density lipoprotein (LDL) receptors from the endosome to the cell surface, after the LDL particles themselves have been internalised for the purposes of removing their cholesterol cargo. The canonical model is that a membrane protrusion forms, thereby excluding soluble proteins. Considering dynamin's remarkable ability to constrict tubes, it is feasible that a dynamin analogue such as the VPS1 protein could facilitate membrane protrusion (and therefore membrane sorting) by a common mechanism. The VPS1 protein and dynamin-1 both have the 'self-assembly' domain essential for the polymerisation of mouse Mx protein,[9] suggesting that VPS1 may also form helices. Hinshaw and Schmid's data, however, caution us that the proline-rich, SH3-binding, carboxy-terminal region, which is not conserved among dynamin homologues, is required for polymerisation of dynamin in vitro.

Nonetheless, the roles of dynamin in membrane fission and of VPS1 in sorting probably both depend on the ability to constrict membrane tubes using helical polymers. If this turns out to be the case, then other mutations that affect sorting should involve dynamin-like proteins.

References
1. Takei, K., McPherson, P.S., Schmid, S.L. & De Camilli, P. *Nature* **374**, 186-190 (1995).
2. Hinshaw, J.E. & Schmid, S.L. *Nature* **374**, 190-192 (1995).
3. Vallee, R.B. & Shpetner, H.S. *Nature* **365**,

107-108 (1993).

4. Robinson, M.S. *Curr. Opin. Cell Biol.* **6**, 538-544 (1994).

5. Carter, L.L., Redelmeier, T.E., Woollenweber, L.A. & Schmid, S.L. *J. Cell Biol.* **120**, 37-45 (1993).

6. Kosaka, T. & Ikeda, K. *J. Neurobiol.* **14**, 207-225 (1983).

7. Koenig, J.H. & Ikeda, K. *J. Neurosci.* **9**, 3844-3860 (1989).

8. Damke, H., Baba, T., Warnock, D.E. & Schmid, S.L. *J. Cell Biol.* **127**, 915-934 (1994).

9. Nakayama, M. *et al. J. Biol. Chem.* **268**, 15033-15038 (1993).

10. Vater, C.A., Raymond. C.K., Ekena, K., Howald-Stevenson, I. & Stevens, T.H. *J. Cell Biol.* **119**, 773-786 (1992).

11. Wilsbach, K. & Payne, G.S. *EMBO J.* **12**, 3049-3059 (1993). ❏

Questions

1. What are synaptic vesicles?

2. What is endocytosis?

3. What role does the protein dynamin play in endocytosis?

Answers are at the back of the book.

7

The plasma membrane of the cell which forms the outer most boundary was first reported by cell biologists more than two decades ago. Its structure was proposed as a double layer of lipid molecules with proteins anchored on either side as well as some panning the entire membrane surface. Other components such as cholesterol and carbohydrates could also be found within the plasma membrane. But when describing its appearance, scientists saw something quite interesting. The proteins were moving about the fluid membrane, changing its presentation like a mosaic pattern, thus it became known as the fluid mosaic model. This model has remained the framework for much of the work on membranes for many years. Current research on membranes using highly sophisticated techniques, however, is revealing new information which challenges some of the old ideas about membrane dynamics and the fluid mosaic model.

Revisiting the Fluid Mosaic Model of Membranes

Ken Jacobson, Erin D. Sheets, and Rudolf Simson

Science, June 9, 1995

The fluid mosaic model, described over 20 years ago, characterized the cell membrane as "a two-dimensional oriented solution of integral proteins...in the viscous phospholipid bilayer."[1] This concept continues as the framework for thinking about the dynamic structure of biomembranes, but certain aspects now need revision. Most membrane proteins do not enjoy the continuous, unrestricted lateral diffusion characteristic of a random, two-dimensional fluid. Instead, proteins diffuse in a more complicated way that indicates considerable lateral heterogeneity in membrane structure, at least on a nanometer scale. Certain proteins are transiently confined to small domains in seemingly undifferentiated membrane regions. Another surprise is that a few membrane proteins undergo rapid, forward-directed transport toward the cell edge, perhaps propelled by cytoskeleton motors.

The more detailed view of the life of a membrane protein has emerged as a result of one old and two newer methods. For the past two decades, fluorescence recovery after photobleaching (FRAP) has been the major tool for measuring the lateral mobility of membrane components labeled directly with fluorophores or with fluorescent antibodies. In this method, a short pulse of intense laser light irreversibly destroys (photobleaches) the fluorophores in a micrometer-sized spot. The florescence gradually returns as fluorophores from the surrounding region diffuse into the irradiated area. FRAP experiments can reveal the fraction of labeled membrane proteins or lipids that can move, the rate of this movement (characterized by the lateral diffusion coefficient), and the fraction of proteins that cannot move on the time scale of the experiment. These apparently nondiffusing proteins are called the immobile fraction; a quantity that is frequently large and usually of unknown origin.

A second method, single particle tracking (SPT), directly complements the information that is obtained from averaging the movement of hundreds to thousands of molecules in a FRAP experiment. In SPT, a membrane component is specifically labeled with an antibody-coated submicrometer colloidal gold or fluorescent particle, and the trajectory of the labeled molecule is followed with nanometer precision with digital imaging microscopy.[2,3] Visualization of individual protein motions can reveal submicroscopic membrane structures as the protein encounters obstacles in its path, although careful data analysis is required to distinguish between nonrandom and random movements.[4]

The third method, recently applied to membranes, is the optical laser trap, allowing further characterization of the obstacles a membrane encounters. Proteins are labeled with submicrometer beads and manipulated in the plane of the membrane with laser light. Optical trapping occurs when a near-infrared laser beam with a bell-shaped intensity profile is focused on the bead attached to the protein. Optical forces on the bead, which are directed toward the highest intensity of the beam, trap the particle.[5] By moving the laser beam or the microscope stage, the labeled protein can be dragged across the plasma membrane until it encounters a barrier or obstacle that causes the bead to escape the trap. The distance between barriers is called the barrier-free path (BFP).

The fluid mosaic model proposes random, two-dimensional diffusion for membrane components. Although lipids[6] and a fraction of the labeled protein population appear by SPT to diffuse freely, other protein movements are considerably more complicated than originally envisioned in the fluid mosaic model. One big surprise has been that a substantial fraction of the proteins are confined, at leas transiently, to small domains. This has been seen most clearly for certain cell adhesion molecules [cadherins and neural cell adhesion molecules (NCAMs)] and nutrient and growth factor receptors. For cadherins, transferrin receptors, and epidermal growth factor receptors, the domains are 300 to 600 nm in diameter and confinement lasts from 3 to 30 s.[7] Following earlier work on the red cell membrane. Kusumi and colleagues[7] proposed the "membrane-skeleton fence" model. In this scheme, a spectrin-like meshwork closely apposed to the cytoplasmic face of the membrane sterically confines membrane-spanning proteins to regions on the order of the cytoskeleton mesh size. Support for this model includes the facts that partial destruction of the cytoskeleton decreases the fraction of confined molecules, and truncation of the cytoplasmic domain leads to less confined diffusion.[8,9]

Can the fence model be supported by other techniques? Enter the laser trap. In a pioneering study, Edidin et al.[10] showed that BFPs for the lipid-linked and the membrane-spanning isoforms of the major histocompatibility antigens were ~1700 and ~600 nm at the 23°C, respectively, and that these values increased with the temperature, indicating the dynamic nature of the barriers. Using weaker trapping forces, Sako and Kusumi[11] could detect even smaller BFPs for the transferrin receptor (~400 nm) which are consistent with the size of domains measured by SPT for both this receptor and E-cadherin.[7,8] The fences appear elastic, because the transferrin receptor rebounds after it strikes barriers,[11] and a small fraction of these receptors seem to be fixed to the underlying cytoskeleton by spring-like tethers.[11]

To permit the long-range diffusion observed by both SPT and FRAP, these barriers must open temporarily, either by disassociation of key molecular constituents of the barriers or by thermally driven local fluctuations of the meshwork-membrane distance. The escape of a given protein in an adjacent domain probably depends on the size of its cytoplasmic moiety, which implies that the effective domain size may protein-dependent.[7,12] The emerging picture is that the immobile fraction of membrane proteins measured by FRAP does not simply represent stationary proteins by rather is some combination of proteins actually tethered to the cytoskeleton and those moving within and between confinement zones.

Is direct trapping by the cortical cytoskeleton the only means of confinement? Probably not. Surprisingly, confinement was also found for a lipid-linked isoform of NCAM in muscle cells, which cannot be directly trapped by the cytoskeletal network. In this case, the membrane domains were ~280 nm in diameter, an the proteins remained in them for about 8 s.[13] The confinement may be the result of interactions with the same or other proteins that are associated with the cytoskeleton. SPT analysis suggests that the proteins in these zones are diffusing through a dense field of obstacles. Presumably, such domains will transiently trap different proteins, although this has not been proven. Other glycosylphosphatidyl inositol-anchored proteins such as Thy-1 also exhibit tightly confined diffusion,[14] possibly because they are sequestered in glycolipid-enriched regions that include caveolae.[15] Such confinement zones could play a significant role in mediating adhesion or in signal transduction by collecting relevant molecules, for example, cell adhesion molecules with their cooperating growth factor receptors.[16]

A diverse set of membrane proteins can also be seen with SPT to move by highly directed, nondiffusional transport, sometimes in unexpected directions. Some proteins go in the direction opposite that of the bulk movement of patches of cross-linked proteins into caps seen in lymphocytes and other cells.[2,17] For example, integrins move outward toward the cell periphery in a highly directed fashion.[18] These cell matrix adhesion receptors, which are important for cell locomotion, may be recycled from the back to the front of the cell by forward-directed cytoskeletal motors.

The plasma membrane presents an intriguing mix of dynamic activities in which components may randomly diffuse, be confined transiently to small domains, or experience highly directed movements. The coexistence of multiple modes of diffusion and directed transport indicates pronounced lateral heterogeneity in the membrane. Key issues remain:

How generally applicable is the membrane-skeleton fence model? Are transmembrane and glycosylphosphatidyl inositol-anchored proteins confined by the same structures in the skeleton? How is the domain structure regulated? The greatest challenge will be to relate this exciting new knowledge of membrane dynamics to the manifold function accomplished by the plasma membrane.

References

1. S.J. Singer and G. L. Nicolson, *Science* **175**, 720 (1972).
2. M. de Brabander *et al.*, *J. Cell Biol.* **112**, 11 (1991); R.J. Cherry, G.N. Georgiou, I.E.G. Morrison, *Biochem. Soc. Trans.* **22**, 781 (1994); M.P. Sheetz, S. Turney, H. Qian, E.L. Elson, *Nature* **340**, 284 (1989).
3. R.N. Ghosh and W.W. Webb, *Biophys. J.* **66**, 1301 (1994).
4. M.J. Saxton, *ibid*, **64**, 1766, (1993).
5. K. Svoboda and S.M. Block, *Annu. Rev. Biophys. Biomol. Struct.* **23**, 247 (1994).
6. G.M. Lee, F. Zhang, A. Ishihara, C.L. McNeil, K.A. Jacobson, *J. Cell Biol.* **120**, 25 (1993).
7. A. Kusumi, Y. Sako, M. Yamamoto, *Biophys. J.* **65**, 2021 (1993).
8. Y. Sako and A. Kusumi, *J. Cell Biol.* **125**, 125 (1994).
9. R.N. Ghosh and W.W. Webb, *BioPhys. J.* **57** 286a (1990); Y. Sako, A. Nagafuchi, M. Takeich, A. Kusumi, *Mol. Biol. Cell* **3**, 219a (1992).
10. M. Edidin, S.C. Kuo, M.P. Sheetz, *Science* **254** 1379 (1991).
11. Y. Sako and A. Kusumi, *J. Cell Biol.*, in press.
12. M. Edidin, M.C. Zuniga, M.P. Sheetz, *Proc. Nat. Acad. Sci. U.S.A.* **91**, 3378 (1994).
13. R. Simson *et al.*, *Biophys. J.* **68**, A436 (1995).
14. E.D. Sheets, G.M. Lee, K. Jacobson, *ibid.*, A306.
15. D.A. Brown and J.K. Rose, *Cell* **68**, 533 (1992) R.G.W. Anderson, *Curr. Opin. Cell Biol.* **5**, 64 (1993).
16. E.J. Williams, J. Furness, F.S. Walsh, P.

Doherty, *Neuron* **13**, 583 (1994); C. O'Brien, *Science* **267** 1263 (1995).

17. B.F. Holifield and K. Jacobson, *J. Cell Sci.* **98** 191 (1991).

18. C.E. Schmidt, A.F. Horwitz, D.A. Lauffenburger, M.P. Sheetz, *J. Cell Biol.* **123**, 977 (1993).

19. We thank a number of colleagues for their helpful comments. ❏

Questions

1. What is the fluid mosaic model?

2. What groups of proteins have been found confined in small domains in the plasma membrane?

3. What does the "membrane-skeleton fence" model propose?

Answers are at the back of the book.

8

Homeostasis is essential for survival of cells, especially since they function only within a very narrow range of composition in the extracellular fluid. The buffer systems in the body contribute to homeostasis by maintaining the proper pH in the internal environment. The reaction of metabolically produced CO_2 with water to make bicarbonate and H^+ is an important reaction which is catalyzed by the enzyme carbonic anhydrase. Because the need to transport bicarbonate across the membrane is great, researchers have focused their efforts on studying the membrane-transport mechanism for bicarbonate. A new transporter, as reported in this article, may be important for the control of intracellular pH regulation. Unlike the well characterized chloride / bicarbonate exchanger, this new system depends on the cotransport of potassium. Its exact physiological function, however, still remains to be determined.

Bicarbonate Briefly CO$_2$-Free

Roger C. Thomas

Nature, April 13, 1995

Animal life is a remarkable complex mechanism for converting oxygen to carbon dioxide. The subsequent reaction of CO_2 with water to make bicarbonate and H^+ gives bicarbonate a central role in pH homeostasis. Numerous different membrane-transport mechanisms for bicarbonate are known, and another one emerges from the ingenious studies with out-of-equilibrium solutions reported by Zhao and colleagues.

The slowness of the reaction of CO_2 with water is overcome in physiology by the enzyme carbonic anhydrase.[2,3] This was famously first found in red blood cells where it maximizes the capacity of blood to carry CO_2 (as bicarbonate). The first bicarbonate transporter recognized was indeed the chloride/bicarbonate exchanger of the same cell. It allows the rapid interchange of bicarbonate and chloride, to generate the chloride shift. Since then many other bicarbonate transporters have been discovered, the

one of most general importance in intracellular pH regulation probably being the Na-dependent Cl^{-1} HCO_3^- exchanger.[4,5]

A similar mechanism regulates pH in squid giant axons[6] but bicarbonate is still carried across the axon membrane in the complete absence of Na^+ or Cl ions.[1] It proved difficult to investigate the mechanism involved using conventional CO_2/HCO_3^- solutions, because the CO_2 makes it impossible to maintain a significant bicarbonate gradient. The normal bicarbonate saline of pH 8 used in squid-axon experiments contains 12 mM bicarbonate and must be equilibrated with 0.5% CO_2 to be stable. If this equilibrium is disturbed, for example by the loss of CO_2 to the atmosphere, the pH will slowly increase as bicarbonate too is converted to CO_2 and lost.

A stable bicarbonate solution must therefore contain CO_2. This crosses cell membranes so fast that its level inside cells is only rarely different from that

Reprinted by permission of *Nature*, Vol. 374, April 13, 1995, pp. 597-598.
Copyright 1995 Macmillan Magazines Ltd.

outside. The intracellular CO_2 is in equilibrium with intracellular H^+ and HCO_3^-, often catalyzed by the almost ubiquitous carbonic anhydrase. So you cannot normally have external bicarbonate without a similar level inside cells. The exact level will depend on the intracellular pH, but his is usually close to the external pH.

Zhao and colleagues[1] solved the problem by the clever use of out-of-equilibrium solutions. They mixed two pairs of carefully formulated solutions in a continuous-flow apparatus which delivered the mixture to the outside of a squid axon within half a second. They mixed either a 24 mM bicarbonate alkaline solution or a 1% CO_2 acid solution with acid or alkaline CO_2/bicarbonate-free solutions to give pH 8 mixtures with 12 mM bicarbonate or 0,5% CO_2. The rate constants of the conversion of bicarbonate to CO_2 and vice versa are such that very little takes place within half a second. All solutions contained acetazolamide to inhibit any carbonic anhydrase that might be present.

That the CO_2-free, freshly mixed solution indeed contained almost no CO_2 is shown convincingly by its intracellular buffering power, which Zhao et al, measured by applying ammonium solutions. The equilibrated CO_2/HCO_3^- solution increased buffering power by 57%, as expected, whereas the out-of-equilibrium solution increased it by only 7%.

In contrast to its lack of effect on buffering, the CO_2-free, out-of-equilibrium bicarbonate solution doubled bicarbonate uptake rates by potassium-depleted axons in high-K solution, strong evidence for cotransport of K and HCO_3^-. Similarly, bicarbonate efflux from K-loaded axons in bicarbonate-free solutions was much faster if the bathing solution contained CO_2. The CO_2 outside it converted to about 2 mM bicarbonate inside, so there is a large outward bicarbonate gradient as well as K gradient to drive the cotransporter.

The physiological function of this hitherto unknown K/HCO_3^- cotransporter is obscure. But the new technique applied by Zhao et al. may breathe fresh life into the study of bicarbonate transport generally.

References

1. Zhao, J., Hogan, E.M., Bevensee, M.O. & Boron, W.F. *Nature* **374**, 636-639 (1995).
2. Maren, T.H. *Physiol. Rev.* **47**, 595-781 (1967).
3. Widdas, W.F., Baker, G.F. & Baker, P. *Cytobios* **80**, 7-24 (1994).
4. Thomas, R.C. *J. Physiol. Lond.* **273**, 317-338 (1977).
5. Schweining, C.J. & Boron, W.F.J. *Physiolo. Lond.* **475**, 59-67 (1994).
6. Boron, W.F. & Russell, J.M. *J. Gen. Physiol.* **81**, 373-399 (1983). ❑

Questions

1. What reaction plays a central role in pH homeostasis in the cell?

2. What enzyme catalyzes the reaction of CO_2 with water?

3. What ion is necessary to drive the cotransporter for bicarbonate in this system?

Answers are at the back of the book.

Section Two

What's New in Tissue Engineering

9   *Call it a medical miracle, but regeneration in the spinal cord which had been thought to be impossible is becoming a reality. Doctors in Sweden and London have been reporting that axons, lost from the spinal cord during injury, have the capability to grow new axons into the muscles if reattached within the first month. If reconnected early, experts say the majority of the cell bodies are still alive as demonstrated in experimental animals. It takes many months for the axons to grow but that is significantly better than a lifetime of disability. Perhaps one day spinal nerve regeneration will even be used to repair severed spinal cords, the most devastating of all spinal injuries.*

Paralysis Lost

Clare Thompson

***New Scientist*, April 6, 1996**

Norrtälje, Sweden. Winter 1993. Twenty-five year old Thomas Westberg makes the fateful decision to take his new Polaris motorbike for a spin. The weather is abysmal. Pouring rain distorts his vision. Icy snow coats the road. But Westberg is still doing over 100km/h, when he skids, and loses control. The accident breaks Westberg's shoulder blade, and—far more ominously—rips several key nerves from their moorings. As Westberg lays crumpled on the ground experiencing the worse pain of his life, he has no idea that less than one year later he will help make medical history.

Westberg had yanked four nerves from where they emerged from his spinal cord in the neck region. The injury left his left shoulder, arm and hand completely paralysed. The prognosis for this damage, which is regarded as a particularly severe form of brachial plexus injury, is not good. The paralysis is considered incurable and permanent, and the arm withers through lake of use.

But through a medical feat that some experts still have difficulty accepting, Westberg regained the use of his are after neurosurgeon Thomas Carlstedt at Karolinska Hospital in Stockholm reattached two of the torn nerves to their original site in the spinal cord. Carlstedt's repair job appears to have provided a channel along which new axons, the projections that carry nerve messages, could grow from the cell bodies in Westberg's spinal cord to the muscles of his arm. Today, Westberg can lift a kilogram weight, and he has regained about 40 percent of normal movement in his arm, which has made the difference between lifetime of disability and continuing in his job as a car mechanic.

His remarkable recovery also challenges the conventional notion that when the nerve cells of the spinal cord are damaged they die, and are neither repaired or replaced, but irretrievably lost. Against all the odds, it appears that cells in the Westberg's spinal cord have grown new axons, which must be well over a foot long and extend into two of his major are muscles.

Not surprisingly, reaction to this "miracle" recovery has been mixed. Neurophysiologist Clifford Woolf of University College London is cautious. He says that the results "strongly suggest" that new axons have grown from the spinal cord to the muscle, but he adds that Carlstedt and his colleagues must successfully repeat the operation in more patients, and ultimately confirm their finding by showing a direct axonal connection from the spinal cord to the muscles in a postmortem examination of those patients.

Ten Tedious Years

Surgeon Rolf Birch at London's Royal Orthopaedic Hospital, on the other hand, is a firm believer. He calls the technique "a truly significant breakthrough" which provides the only hope of restoring function in many brachial plexus injuries. Carlstedt has just completed a one-month stint in London working with Birch and his colleagues, who want to develop the technique to treat their share of the 350 British adults who suffer brachial plexus injuries each year, usually in car, motorbike and water-skiing accidents.

Getting to the point where other neurosurgeons were prepared to take seriously Carlstedt's idea for the brachial plexus repair operation was a long haul. It too "ten long, tedious years," before he had perfected is operation in animals, he says. That work showed that the cell bodies in the spinal cord start to die within two weeks of the spinal nerves being damaged. But if the nerves were reconnected during the first month while the majority of the cell bodies were still alive, new axons would grow to the muscles.

Carlstedt's decision to attempt the procedure in humans coincided with Westberg's accident and, mindful of the time factor, he had him under the knife within weeks. To expose the spinal cord, Carlstedt cut through the bony spinal column and the thick protective membrane called the dura. Two of the four spinal nerves which had been damaged in the accident were completely wrenched from the

spinal column and were beyond repair. However, the other two were pulled from the cord, but not the column, and so were candidates for reconnection.

Spinal nerves fork just outside the spinal cord. One root carries sensory information to the spinal cord from the receptors in the skin, organs and muscles; the second carries instructions to the muscles about when to contract. Carlstedt took the second root of each nerve and surgically attached it just beneath the surface of the cord, in one case using a graft taken from a minor nerve in the leg as an extension.

Axons grow only 1 millimetre per day, so it too many months for them to grow from the spinal cord to the muscles of the arm. Nine months after the surgery, Carlstedt first recorded electrical activity that was characteristic of a nervous input in the biceps and triceps of Westberg's left arm, suggesting that cell bodies in the spinal cord had sent new axons to the muscles. Shortly after that, Westberg found that he was regaining the ability to contract those muscles.

In order to eliminate the possibility that another nerve had somehow compensated for the damaged ones, Carlstedt injected a local aneasthetic directly into the intact nerve that was most likely to have taken over some of the function of the damaged ones. Blocking that nerve hampered the deltoid muscle on top of Westberg's left shoulder, but had no effect on his new-found ability to contract his biceps and triceps (*Lancet*, vol 346, p 1323).

Carlstedt admits that, at first, even he found it hard to believe Westberg's spectacular recovery despite the fact that his animal studies had strongly suggested that the surgery would work. In a key experiment, he had performed the same operation on six monkeys with similar spinal injuries. Three regained the use of their damaged arms. In postmortem examinations of these animals, Carlstedt found that a dye injected into the monkeys' biceps before their death had diffused along axons connecting the muscle to the spinal cord and ended up in cell bodies within the spinal cord. This proved that cell bodies had sent out new axons that went all the way to the muscles.

Since Westberg's recovery, Carlstedt has restored muscle function to another patient and is waiting to see what happens to four others who have undergone the operation. Two more patients have showed no improvement. Carlstedt suspects that the problem with these patients is that their operations were delayed for more than 4 months as they were referred to the Karolinska from other hospitals.

Now Carlstedt and Morten Reisling, a neuroscientist at the Karolinska Hospital, are looking for ways to further improve movement in the arm after the operation, and to widen the window of time for successful surgery. When axons outside the spinal cord or the brain are damaged, chemicals called neurotrophic factors stimulate their regrowth. Earlier this year, neurobiologist Fred Gage at the Scripps Institute in La Jolla, California, announced that implanting connective tissue cells that make neurotrophic factors into the brains of mice also stimulates the growth of new axons. Carlstedt speculates that when the spinal nerves are reattached they mimic that effect by sucking up neurotrophic factors from outside and transporting them to the spinal cord. Carlstedt and Reisling are now testing whether injecting these factors can also help keep cell bodies alive in the damaged spinal cords of rats.

Carlstedt is confident that his spinal nerve operation can be refined. Perhaps one day it will even be used to help repair severed spinal cords, the most devastating of all spinal injuries. After all, the major impediment has been overcome, he says. "We've shown that regeneration in the spinal cord, which had been thought to be impossible, can happen." ❏

Questions

1. How is spinal cord regeneration possible?

2. How fast do axons grow?

3. What natural compound can stimulate axon regrowth?

Answers are at the back of the book.

10

Fibrin sealant, composed of fibrinogen and thrombin, is the most successful and widely used tissue sealant available. It is a "natural" glue that promotes clot formation, and is used routinely in many types of cardiac and vascular surgery to reduce blood loss. The US Army is interested in developing the product into a ready-to-use dressing for the treatment of bleeding combat casualties. In addition to its use as a homeostatic agent, fibrin sealant is considered a suitable matrix for the delivery of antibiotic drugs and other biological agents like growth factors. Perhaps the greatest importance is the improved survival and shorter hospital stays for patients whose wound healing could benefit from fibrin sealants.

Tissue Sealants: Current Status, Future Potential

Mark R. Jackson

Nature Medicine, May 1996

Fibrin sealant has clot-promoting activity that halts bleeding from injury or during surgery.

Fibrin sealant or fibrin "glue" is a tissue sealant which mimics the final step of the blood coagulation cascade. The fibrinogen and thrombin components of fibrin sealant are combined using a dual—syringe system similar to that used to prepare epoxy glue from the separate resin and hardener. While several different sealants are in various stages of research and development, fibrin sealant has been the most successful and widely used tissue sealant to date. Fibrin sealant is used clinically as an adjunct to hemostasis to aid in the arrest of blood loss during surgery (particularly cardiac and vascular surgery) and to halt blood loss in the management of trauma patients. In 1978 the Food and Drug Administration (FDA) revoked the license for commercial fibrinogen concentrates (because of the high rates of

hepatitis resulting from virally—contaminated blood used in fibrinogen preparation.[1] and while commercial development continued in Europe, it was abandoned in the United States. Careful donor selection and heat treatment as well as other viral inactivation methods, have now largely resolved the issue of viral safety. Renewed efforts to develop a commercially—available fibrin sealant product for use in the United States are underway and were the primary focus of a recent meeting in La Jolla, California,* convened to discuss recent research advances and regulatory issues related to tissue sealants.

The fibrin sealant used in many U.S. medical centers is prepared from donor cryoprecipitated blood plasma as the source of fibrinogen (with careful donor selection and screening to address issues of viral safety) and bovine thrombin. However, some

* *Tissue sealants: current practice, future uses.* March 20-April 3, La Jolla, California.

Reprinted with permission from *Nature Medicine*, Vol. 2, No. 5, May 1996.

commercial preparations of bovine thrombin contain high concentrations of bovine factor V. Antibodies against bovine factor V (which cross-react with human factor V) have been detected in human patients treated with fibrin sealant and have resulted in factor V deficiency of sufficient severity to cause bleeding complications.[2,3] Commercial production of fibrin sealant prepared from purified human thrombin and human fibrinogen should circumvent these problems.

The advantages of using fibrin sealant to arrest bleeding in cardiac, vascular and other types of surgery have been shown in a number of studies. In a randomized, prospective, multicenter U.S. trial which evaluated a commercially—prepared fibrin sealant for arresting blood loss during cardiac surgery, successful hemostasis was achieved within five minutes in 92.6% of those treated with fibrin sealant compared with only 12.4% in those treated with conventional topical hemostatic agents.[4] An intent—to—treat analysis performed retrospectively (on the prospectively acquired data) showed improved survival and a shorter hospital stay in the group treated with fibrin sealant.[5] A recent clinical trial evaluated fibrin sealant (prepared by the Scottish National Blood Transfusion Service) to prevent bleeding through suture holes in patients undergoing carotid endarterectomy.[6] This is a commonly performed surgical procedure that involves removal of obstructing plaque from the major artery supplying blood to the brain thereby reducing the risk of stroke. In the Scottish study, a patch of polytetrafluoroethylene was used to close the carotid artery after surgery and patients were randomized to receive either a patch treated with fibrin sealant or a untreated patch. Patients receiving the treated patch had a statistically significant reduction in blood loss and a shorter time to hemostasis.

At the California meeting investigators from the Walter Reed Army Institute of Research and the American Red Cross presented data on the hemostatic efficacy of a newly designed fibrin—based dressing in a porcine model of femoral artery injury.[7] In this study, a potentially field-ready fibrin dressing, when applied to experimentally injured arteries, resulted in a statistically significant reduc-

tion in blood loss compared with a visually—identical placebo control dressing. Furthermore, the fibrin dressing caused no reduction in femoral artery blood flow distal to the injury. Investigators from William Beaumont Army Medical Center are also performing preclinical studies to test efficacy in field situations. The U.S. Army has significant interest in developing fibrin sealant in the form of a ready—to—use dressing for the treatment of bleeding combat casualties. TachoComb™, a commercially prepared dressing composed of collagen fleece coated with fibrin sealant, (manufactured by Hafslund Nycomed Pharma AG, Vienna, Austria) has been successfully used in a variety of surgical procedures in randomized (partly uncontrolled) clinical trials involving 2592 patients in Europe.[8] TachoComb™ showed superior hemostasis in a variety of clinical applications compared with either fibrin sealant alone or collagen fleece alone and was well tolerated by patients.

New developments with other tissue sealants were also discussed at the meeting. A composite adhesive of fibrinogen and collagen developed by the Cohesion Corporation (Palo Alto, California) appears to have hemostatic efficacy in animal models as well as good adhesive and mechanical strength.[9] Other collagen based sealants are under development by Cohesion Corporation and Biomatech (Paris, France) and preclinical studies look promising.

In addition to its use as a hemostatic agent, a number of presentations focused on the future applications of fibrin sealant as a matrix for the delivery of drugs and biological agents to the wound site. Groups from Kent State University and the American Red Cross presented data showing that fibrin sealant can be impregnated with antibiotics and that high concentrations can be safely delivered to the site of infection in animal models. The doses of antibiotic delivered by conventional routes are too low to treat the target organisms effectively.[10] Fibrin sealant has also been shown to be an effective delivery matrix for biologic agents such as growth factors. Researchers from Loyola University Medical Center presented data supporting enhanced growth of endothelial cells *in vitro* in prosthetic vascular graft material containing fibroblast growth factor in fibrin sealant.[11] If improved growth of endothelium can be supported in such a graft, this would have significant potential in vascular bypass surgery for ischemia of the legs, where the use of prosthetic graft material has met with limited success given the inability of such grafts to develop an inner layer of endothelial cells.

Product standardization issues were discussed at a roundtable forum. Specifically, optimal concentrations of the major components of fibrin sealant (fibrinogen and thrombin) and whether to include other adjunctive components (such as the antifibrinolytic agent, aprotinin) generated considerable controversy and little consensus. It is possible that different fibrin sealant products will emerge, each with different clinical applications and formulations.

Perhaps of greatest importance from the regulatory perspective were discussions of the proposed *potential* revisions to clinical trial assessment by the FDA. Whereas the existing guidelines for clinical trials assessment require demonstration of efficacy with hard clinical endpoints such as in bleeding complications which require additional surgical treatment or a blood transfusion, the *potential* revision might allow simply for the demonstration of improvements in hemostasis. Date might be compared with placebo or an approved product, or with the current standard of care, on the understanding that the appropriateness of the chosen control must be justified. This marks a potentially significant departure from existing policy, which if enacted, could expedite the licensing of commercial preparations of fibrin sealant.

Purified, human fibrin sealant products have undergone considerable scientific development with demonstrated viral safety and evidence supporting hemostatic effectiveness. These products are likely to be commercially available in the U.S. for certain specific indications in the not—too—distant future. In addition to their use as an adjunct to hemostasis in the operating room and on the battlefield, these materials have exciting potential as delivery vehicles for drugs and other biological materials in the treatments of a wide variety of medical and surgical disorders.

The opinions expressed herein are the private views of the author and are not to be construed as official or reflecting the views of the Army or the Department of Defense.

References

1. Alving, B.M., Weinstein, M.J., Finlayson, J.S., Menitove, J.E. & Fratantoni, J.C. Fibrin sealant: Summary of a conference on characteristics and clinical uses. *Transfusion* **35**, 783-790 (1995).
2. Zehnder, J.L. & Leung, L.L.K. Development of antibodies to thrombin and factor V with recurrent bleeding in a patient exposed to topical bovine thrombin. *Blood* **76**, 2011-2016 (1990).
3. Nichols, W.L. et al. Antibodies to bovine thrombin and coagulation factor V associated with surgical use of topical thrombin or fibrin "glue:" A frequent finding. *Blood* **82**, 59 (1993).
4. Rousou, J. et al. Randomized clinical trial of fibrin sealant in patients undergoing resternotomy or reoperation after cardiac operations: A multicenter study. *J. Thorac. Cardiovasc. Surg.* **97**, 194-203 (1989).
5. Levitsky, S. Randomised clinical trial of Immuno's fibrin sealant in patients undergoing resternotomy or reoperation after cardiac operations: A multicenter study. *Fibrin Sealant: Characteristics and Clinical Uses.* (abst.) (1994).
6. Milne, A.A., Murphy, W.G., Reading, S.J. & Ruckley, C.V. Fibrin sealant reduces suture line bleeding during carotid endarterectomy; a randomised trial. *Eur. J. Vasc. Endovasc. Surg.* **10**, 91-94 (1995).
7. Jackson, M.R. et al. Randomized blinded comparison of a fibrin-based dressing with a placebo control in a femoral artery injury model. *Cambridge Symposia* **1**, 26 (abst.) (1996).
8. Engelsen, S.J., Kuntz, G., Homdrum, E.M. & Kuhlberger, E. A new approach in the management of intraoperative bleeding: The ready-to-use combination of fibrin sealant with a collagen fleece. *Cambridge Symposia* **1**, 26 (abst.) (1996).
9. Sierra, D.H., Kloss, J., Prior, J. & Wilson, D. Collagen-fibrinogen composite tissue adhesive.

Cambridge Symposia **1**, 24 (abst) (1996).

10. Woolverton, C.J., Singh, M., Drohan, W.N. & MacPhee, M.J. Long-term antibiotic release from fibrin sealant. *Cambridge Symposia* **1**, 30

(abst.) (1996).

11. Greisler, H.P. Neovascularization of ePTFE peripheral vascular grafts. *Cambridge Symposia* **1**, 28 (abst.) (1996). ❏

Questions

1. What are the biological components of fibrin?

2. Since fibrin is a blood component, what safety precautions must be taken in preparation for its use as a sealant?

3. In addition to its use as a homeostatic agent, what other applications are being considered for fibrin sealants?

Answers are at the back of the book.

11 *Scientists have been puzzled for years why the central nervous system does not support neural growth to the same extent as its peripheral cousin. Conditions for peripheral nerve grafts can be successful if those nerves grow in a favorable environment, however, some factors appear to be inhibiting the central neurons from regenerating. One such factor found in higher concentrations in the myelinated axons of central neurons is a myelin-associated glycoprotein (MAG) which appears to prevent the collateral sprouting of established neural connections. In addition to MAG, researchers have now uncovered a host of other factors with growth-inhibitory actions. Although the nerve regeneration puzzle has many more pieces to solve, the future for those with brain and spinal cord injuries is brighter.*

Making the Connections in Nerve Regeneration

Julia Nash and Adrian Pini

***Nature Medicine,* January 1996**

New findings suggest that there are multiple inhibitory signals for regenerating neurons.

You don't have to be a neuroscientist or even a rocket scientist to appreciate the terrible consequences of injury to the brain or spinal cord. The neurological picture in peripheral nerve injury is little better, although the outcome for the patient may be less serious. As de Medinaceli and Seaber[1] note, "Despite the application of meticulous technique to peripheral nerve repair, no adult with a major peripheral nerve transection has ever attained normal sensibility." To date, research into the effects of axonal injury has largely been directed along two lines. The first, and by far the least well understood, concerns the immediate intrinsic reaction of neurons to axotomy. The second approach centers on the longer term effects mediated by the environment that regenerating axons encounter.

Despite the prescient contributions of Tello and Ramón y Cajal (the forefathers of neuroscience), the prevailing view up to the 1980s was that neurons of the mammalian central nervous system did not regenerate following injury. This dogma came to grief with the landmark demonstration of David and Aguayo[2] that severed central axons could be induced to regrow through the conduit of peripheral nerve grafts. Although spectacular, it was pretty clear (even at this early stage) that only a small proportion of injured axons regrew. The questions of why this is so and to what extent functional connections can be remade are still largely unresolved.

Aguayo and his colleagues pioneered transection of the optic nerve as a paradigm to study the effects of injury within the central nervous system. When peripheral nerves were grafted to the severed adult optic stalk approximately 10 percent of retinal ganglion cell axons regrew[3] and, remarkably, some of these made synaptic connections with their targets in the superior colliculus.[3] Thinking was firmly in favor of the idea that, given suitable conditions, severed axons could regrow, but were prevented from doing so because of an unfavorable central environment. Support for this view came from observations that optic nerve transection alone resulted in very rapid loss of virtually all retinal ganglion cells.[4]

Within peripheral nerve grafts the most likely candidates for promoting survival and regeneration are components of the basal lamina of Schwann cells, the extracellular matrix, and growth factors. Why, then, is the central nervous system so unfavorable to regeneration? To attempt an answer to this question, Schwab and his colleagues have partially purified central myelin-derived protein fractions that inhibit axonal extension *in vitro*.[5] *In vivo*, treatment with a neutralizing antibody (IN-1) raised against these fractions promotes axonal extension.[6] These results are tantalizing, but again, as seen with peripheral nerve grafts, the proportion of injured axons that regrow is similarly small. This raises the questions as to whether there are additional inhibitory activities (in central myelin or elsewhere) and of the extent to which peripheral myelin impedes regeneration.

Further developments came with the surprising findings of two groups that myelin-associated glycoprotein (MAG) can inhibit axonal growth *in vitro*.[7,8] As had been expected,[9] MAG promoted the outgrowth of neonatal dorsal root ganglion cells but inhibited that of embryonic cerebellar neurons, adult dorsal root ganglion cells, and a motor neuron-like cell line. In myelinated axons of central and peripheral neurons MAG is localized to the innermost myelin membrane in contact with the axon. However, in peripheral nerves (where its abundance is one-tenth of that found centrally) its distribution also includes the outer myelin loops and interstitial areas. Although its function is not yet clear, it is possible that MAG may inhibit regeneration by way of a normal function that prevents the collateral sprouting of established connections.

MAG is clearly distinct from the proteins recognized by the IN-1 antibody, and so represents a second myelin-derived inhibitory activity. However, the results of two very recent experiments in which the regenerative capacities of MAG knockout mice were examined are a variance on this point. Once group has shown improved peripheral regeneration,[10] whereas another finds no effect centrally.[11] The fact that MAG levels in peripheral nerve are relatively low could be seen as consistent with earlier results in which embryonic neurons were grown successfully on ultrathin sections of peripheral nerve.[12,13] Morever, purified peripheral myelin preparations appeared be good substrates for axon growth.[6]

But then Bedi *et al.* upset the apple cart by demonstrating that adult dorsal root ganglion cells do not grow on thin sections of intact peripheral nerve, but only on those form which myelin has been removed by Wallerian degeneration.[14] Now, David *et al.* make an important contribution by directly demonstrating that peripheral nerve myelin is inhibitory to the growth of neonatal neurons.[15] But there's a twist: They prepared myelin from adult bovine peripheral nerves using a homogenizer which operates at low shear forces. The method used previously,[5] which gave a myelin preparation that supported neural growth, employed very high shear forces. David and collaborators reasoned that the discrepancy between the two preparations might have resulted from the presence of laminin in myelin prepared under the latter conditions; at high shear forces laminin (which is enriched in the basal lamina) may have liberated and copurified with the myelin. Western blots revealed that this was indeed the case, and that the growth-promoting activity in myelin prepared in this way could be blocked by antibodies to laminin. Moreover, laminin added peripheral (or central) myelin prepared at low shear forces overrides its growth-inhibitory properties, an observation which may suggest a therapeutic approach.

It seems then, as Bedi et al. originally suggested, that the growth of adult dorsal root ganglion cells on degenerated peripheral nerve is supported by laminin and inhibited in intact nerves by myelin. But why, as they also showed, do embryonic dorsal root ganglion cells grow on intact peripheral nerves? and why, as Shewan et al.[16] have shown, do neonatal retinal ganglion cells and embryonic and dorsal root ganglion cells all fail to grow on unmyelinated sections of perinatal optic nerve? To further complicate things, Keynes and his colleagues have isolated a glycoprotein fraction from adult gray matter that inhibits growth of both central and peripheral neurons *in vitro*.[17] In addition, there are the growth-inhibitory actions of the collapsins/semaphorins, netrins, and ligands of the Eph receptor tyrosine kinases such as RAGS.[18]

So where does all this lead us? The picture is getting complicated and clearly points to multiple stop signals (including those associated with myelin). The implication is that no single strategy will entirely defeat the unregenerate nervous system. Recently,[19] Anders Björklund quoted Ralph Gerard, who some 40 years previously said: "The exciting thing, I repeat, is that once one knows regeneration is inherently possible it is just a question of making it work." It looks like one hell of a job—but one that might well be possible.

References

1. de Medinaceli, L. & Seaber, A.V. Experimental nerve reconnection: importance of initial repair. *Microsurgery* **10**, 56-70 (1989).
2. David, S. & Aguayo, A.J. Axonal elongation into peripheral nervous system "bridges" after central nervous system injury in adult rats. *Science* **214**, 931-933 (1981).
3. Kierstead, S.A. et al. Electrophysiologic responses in hamster superior colliculus evoked by regenerating retinal axons. *Science* **246**, 255-257 (1989).
4. Villegas-Perez, M-P., Vidal-Sanz, M., Rasminsky, M., Bray, G.M. & Aguayo, A.J. Rapid and protracted phases of retinal ganglion cell loss follow axotomy in the optic nerve of adult rats. *J. Neurobiol.* **24**, 23-36 (1993).
5. Caroni, P. & Scwab, M.E. Two membrane protein fractions from rat central myelin with inhibitory properties for neurite growth and fibroblast spreading. *J. Cell Biol.* **106**, 1281-1288 (1988).
6. Schnell, L., Schneider, R., Kolbeck, R., Barde, Y-A. & Schwab, M.E. Neurotrophin-3 enhances sprouting of corticospinal tract during development and after adult spinal cord lesion. *Nature* **367**, 170-173 (1994).
7. McKerracher, L. et al. Identification of myelin-associated glycoprotein as a major myelin-derived inhibitor of neurite growth. *Neuron* **13**, 805-811 (1994).
8. Mukhopadhyay, G., Doherty, P., Walsh, F.S., Crocker, P.R. & Filbin, M.T. A novel role for myelin-associated glycoprotein as an inhibitor of axonal regeneration. *Neuron* **13**, 757-767 (1994).
9. Johnson, P.W. et al. Recombinant myelin-associated glycoprotein confers neural adhesion and neurite outgrowth function. *Neuron* **3**, 377-385, (1989).
10. Schäfer, M., Fruttiger, M., Montag, D., Schachner, M. & Martini, R. Disruption of the myelin- associated glycoprotein (MAG) gene improves axonal regeneration in C57BL/WLD[5] mice. *Abst. Soc. Neuroscience.* **612**.1 (1995).
11. Bartsch, U. et al. Lack of evidence that the myelin-associated glycoprotein (MAG) is a major inhibitor of axonal regeneration. *Abstr. Soc. Neuroscience.* **612**.2 (1995).
12. Carbonetto, S., Evans, D. & Cochard, P. Nerve fiber growth in culture on tissue substrates from central and peripheral nervous system. *J. Neurosci.* **7**, 610-620 (1987).
13. Savio, T. & Schwag, M.E. Rat CNS white matter, but not gray matter, is nonpermissive for neuronal cell adhesion and fibre outgrowth. *J. Neursci.* **9**, 1126-1133 (1989).
14. Bedi, K.S., Winter, J., Berry M. & Cohen J. Adult rat dorsal root ganglion neurons extend neurites on predegenerated but not on normal peripheral nerves *in vitro. Eur. J. Neurosci.* **4**, 193-200 (1992).
15. David, S., Braun, P.E., Jackson, D.L., Kottis, V. & McKerracher, L. Laminin overrides the

inhibitory effects of PNS and CNS myelin-derived inhibitors of neurite growth. *J. Neurosci. Res,* 42, 594-602 (1995)

16. Shewan, D., Berry M., Bedi, K. & Cohen, J. Embryonic optic nerve tissue fails to support neurite outgrowth by central and peripheral neurons in vitro. *Eur. J. Neurosci.* **5**, 809-817 (1993).

17. Keynes, R.J., Johnson, A.R., Picart, C.J., Dunin-Borrowski, O.M. & Cook, G.M.W. A glycoprotein fraction from adult chicken gray matter causes collapse of CNS and PNS growth cones in vitro. *Abstr. Soc. Neurosci.* **16** 77.6 (1990).

18. Keynes, R.J. & Cook G.M.W. Axon guidance molecules. *Cell* 83, 161-169, (1995).

19. Björklund, A. A question of making it work. *Nature* **367**, 112-113 (1994). ❏

Questions

1. What types of injuries could benefit from central nerve regeneration?

2. What factors promote the regeneration of peripheral nerve grafts?

3. What factors may be responsible for inhibiting the central neurons from regenerating?

Answers are at the back of the book.

12 *Consider it hybrid technology, but designer tissue could become a reality in the not so distant future. Researchers are developing laboratory-grown versions of tissues by attaching specific cellular recognition molecules to synthetic polymer surfaces. One synthetic surface that is popular among engineers is Teflon fabric mesh to which protein laminin binds and attracts nerve cells. The technique also shows promise in the growth of liver hepatocytes and blood vessels that have been damaged by disease. Before any implanted material is released, researchers must convince officials that the designer tissue devices are safe.*

Designer Tissues Take Hold

Robert F. Service

Science, October 13, 1995

By engineering polymers that attract and bind liver, nerve, and other cells, scientists are beginning to create artificial organs that are biologically "real."

There's nothing simple about the liver. The organ's many different cell types are arranged in precise three-dimensional patterns to filter toxins from the blood, convert nutrients into forms usable by body tissues, and perform a broad range of other functions. Now scientists are taking basic steps toward growing this complicated organ in the lab.

Tissue engineering, as this field is known, has been around for a little more than a decade. Its researchers have already developed laboratory-grown versions of tissues such as skin and cartilage that are now being tested a replacements for damaged tissues. These are, however, comparatively simple boy parts, consisting of one or two cell types layered on a synthetic polymer scaffold or a mesh made of the structural protein collagen.

A lab-grown liver, however, calls for a new level of sophistication and control: A variety of cells need to be arranged and grown in particular orientations. That means a scaffold with a high degree of selectivity, something polymer and collagen scaffolds lack. But by attaching specific cellular recognition molecules, such as protein fragments, to synthetic scaffolds, biologists and chemists have been able to demonstrate that such feats are indeed possible.

Linda Griffith-Cima, a chemical engineer at the Massachusetts Institute of Technology (MIT), and her colleagues have designed a scaffold that attracts liver cells called hepatocytes while rejecting other cell types. Scientists are using similar hybrid technology to direct the growth of nerve cells on biocompatible materials in hopes of eventually repairing damaged nerves, and to create synthetic blood vessels lined with cells that minimize the formation of dangerous blockages.

Although the work on these hybrid scaffolds is still of a preliminary nature, "it's definitely where the field is going," says David Mooney, an assistant professor of chemical engineering and dentistry at the University of Michigan. The dual nature of the devices is allowing researchers to take advantage of the ability to precisely control the structure of synthetic materials while at the same time camouflaging their foreign nature, minimizing conflicts with the body's immune system. "We not only remove the foreign surface, but we replace it with one that promotes directed growth of the tissue we're interested in," says David Clapper, a cell biologist at BSI Corp., an Eden Prairie, Minnesota-based company working to improve the performance of blood vessels and other implants.

Although the destination is exciting, the field still has to negotiate some bumps before it gets there. One possible barrier, points out Martin Yarmush, a bioengineer at Rutgers University in New Jersey, is that once cells attach themselves to a matrix, hybrid or otherwise, they begin to excrete proteins which may interfere with the carefully laid plans by binding unwanted cells in that region.

Liver under Construction

Selection of specific cell types is a crucial issue in tissue design. Natural and synthetic scaffolds for body tissues, by themselves, are not terribly picky hosts. "Synthetic scaffolding materials are easy and cheap to make and form into a desired structure," says Mooney. But they typically provide holdfasts for many different types of cells, he adds; collagen is similarly undiscriminating. And, notes Mooney, collagen is "hard to isolate in large quantities." So researchers have tried to take advantage of the mass-production capacity of synthetics, but modify them with a biological ability to select specific cell types.

The efforts to bind liver cells illustrate the potential power of such a combination. Thus far, most researches have been trying to design polymers to selectively bind hepatocytes, the most common and important cell type in the liver. These cells carry out more metabolic functions than any other group of cells in the body. One of their duties is to remove damaged proteins from the blood. The cells can do this because they recognize specific carbohydrate sequences, attached to these proteins, that mark these complexes as damaged goods. In the late 1970s Paul Weigel and his colleagues at Johns Hopkins University made synthetic versions of the carbohydrate sequences and used them as lures for the cells:

The scientists attached the carbohydrate to a polymer surface made from polyacrylamide, and saw that the hepatocytes engulfed the lures and remained stuck to the polymer surface.

Although successful at binding hepatocytes, polyacrylamide isn't a viable scaffold for tissue engineering because the material provokes a strong immune response in the body. So more recently Griffith-Cima and her colleagues decided to try attaching the same carbohydrates to a more biocompatible synthetic surface. But they had to choose their material carefully, as many synthetics, such as polypropylene, can be tolerated by the body but quickly become coated with proteins which attract all kinds of cells—exactly what the scientists didn't want.

The researchers decided to build their scaffold with a meshed, water-filled network of polyethylene oxide, or PEO, which is resistant to protein adsorption. PEO molecules are star-shaped, with several arms emanating from a central core. When linked in network in a water-based solution, the end of each arm floats free.

These ends contain reactive hydroxyl groups, to which the researches attached carbohydrate molecules as lures for hepatocyted. Griffith-Cima and her colleagues report in the journal *Biomaterials* (in press) that when they added rat hepatocytes, the cell went right for the lures and ended up bound to the polymer mesh. But other cells, such as fibroblasts, failed to bind to the mesh when Griffith-Cima's group added them to solution. "[Griffith-Cima] has solved one of the big problems—getting a receptor that is unique for hepatocytes," says Kevin Healey, a biomedical engineer at the Chicago campus of Northwester University.

The next step is to build cell-specific scaffolds in three dimensions. A normal liver, which has the largest volume of any organ in the body, consists of a mass of cells tunneled through with blood vessels. The organ's size enables it to filter toxins continuously from the large volume of blood in the body as well as perform its myriad of other functions. To do so, however, the variety of liver cells have to be precisely interspersed. Griffith-Cima's group is taking on these 3D challenges as well. The researchers recently devised a way to use computer-controlled printing techniques to lay down polymers known as polylactic acid (PLA) in specific patterns, one paper-thin layer at a time. That allows them to build up a porous 3D scaffold with a precisely controlled architecture.

The researchers are now working to attach PEO molecules (with the carbohydrate lures) to their PLA scaffold. And over the next couple of years, says Griffith-Cima, they hope to attach other recognition molecules, such as antibodies, to the PEO arms in order to fix liver cells called bile duct cells. They also plan to use the amino acid sequence arginine, glutamic acid, aspartic acid, and valine—known as the REDV sequence—to specifically attract endothelial cells. Eventually, with the proper lures, they hope to fix the right mix of cells for a functioning lever.

Gathering Nerve

Hybrid scaffolds are also useful because they can influence the direction of cell growth. At the State University of New York, Buffalo, Joseph Gardella and his colleagues are using the technique on nerve cells. Ultimately, they and other researchers hope to direct growing neurites—the cell arms that carry nerve impulses—across gaps caused by accidents or illness in the central and peripheral nervous systems. One possibility now being explored by Gardella and his colleagues involves organizing cellular adhesion molecules on a polymer surface.

Gardella, John Ranieri—who recently moved from Brown University to Carbo-Medics in Austin, Texas—and a group of Swiss colleagues led by Patrick Aebischer at the Lausanne University Medical School and Hans Mathieu at the Ecole Polytechnique Fédérale de Lausanne, start with a Teflon fabric mesh. Teflon is considered safe for a variety of biological implants. The researchers then prepare a pattern on it to guide neurites, using a segment of the extracellular matrix protein laminin, which binds to nerve cells as a guide. They begin by covering the polymer mesh with a mask made of nickel with tiny slits cut in the metal. The researchers then expose the nickel-masked Teflon to a hot ionized gas, which penetrates the slits in the nickel and converts fluorine atoms in the fabric to reactive

hydroxyl groups. These hydroxyl groups then act as links to which the researchers attach peptide sequences—known at YIGSR sequences—from laminin.

In the December issue of the *International Journal of Developmental Neuroscience*, the researchers reported that when they placed the mouse nerve cells on their scaffold, the cells bound selectively to regions with the YIGSR-modified Teflon. Moreover, when new neurite branches grew from these cells, they followed the patterned surface. "This is a critical demonstration that you can pattern polymers and nerves will follow the patterns," says MIT's Christine Schmidt, who is researching directed nerve-cell growth. Now the researchers are working to roll their modified fabrics into tube that can be wrapped around damaged nerves in the body.

Designer Liners

Teflon has a long history as another type of implant: artificial blood vessels. But here its history is somewhat spotty. Although the synthetic works well on large-diameter vessels—wider than 6 mm—smaller vessels develop problems. These vessels typically clog up within 2 years after implantation—platelets and smooth muscle cells in the blood begin sticking to the surface of the polymer mesh, occluding the opening. This wouldn't happen if the implant walls more closely resembled natural blood vessels, so scientists are re-engineering them to do just that.

"We're trying to take surfaces that are recognized [by the body] as foreign and change them into something the body really likes," say BSI's Clapper. What the body likes in the case of blood vessel inner walls are endothelial cells, which normally create a slippery surface on those walls that prevents platelets and smooth muscle cells from adhering. The strategy Clapper and several other researchers are pursuing is to induce those cells to bind to the inner walls of polymer vessels.

Researchers have long known that extracellular matrix proteins such as fibronectin and laminin promote endothelial cell binding to different surfaces. Since the early 1980s researchers have identified a number of different peptide sequences of these proteins that are responsible for the adhesion. One, the REDV sequence, was singled out in 1986 by Martin Humphries and Kenneth Yamada at the National Institutes of Health in Bethesda, Maryland. And in 1991 Jeffrey Hubbell, a chemical engineer at the California Institute of Technology in Pasadena, showed that in vitro, the sequence enhances endothelial cell binding to the common graft polymers PTFE and polyethyleneterephthalate.

At the same time work has also progressed with polymers modified with other peptides. In the July issue of the journal *Heart Valve Disease*, researchers report the first in vivo results on a polymer coated with the RGD sequence, made up of arginine, glycine, and aspartic acid. Catherine Tweden and her colleagues at St. Jude Medical, an implant company in St. Paul, Minnesota, along with William Craig and collaborators at Integra Life Sciences in La Jolla, California, report that they implanted RGD-coated polymer patches in the aortas of dogs. After 33 weeks, endothelial cells covered 75% of the RGD coated patches, three times more area than was covered in controls without the peptide coating. The Integra researchers are now implanting RGD-coated synthetic vessels into animals to see if grafts show the same benefit.

The blood vessels, researchers hope, are harbingers of implants to come. Implant researchers are modifying surfaces of a host of other devices, including those for hip joints and breast and dental implants. But like the research on assembling cells into complex tissues, this work remains in its earliest stages. Most of the promising results come from lab studies of how cells interact with hybrid scaffolds. In large part, it remains to be seen how such materials will behave in the bodies of animals, let alone humans. And before that final step can be taken, researchers must convince health officials that any implanted material and its byproducts are safe. Says Rutger's Yarmush, "a lot of detailed work needs to be done." ❏

Questions

1. How do scientists design tissues that are biologically "real"?

2. In using synthetic materials, what are the safety considerations?

3. What are some of the applications for this hybrid technology?

Answers are at the back of the book.

13 *With all the scientific and political controversy surrounding fetal tissue transplants for Parkinson's disease patients, the search for better treatment continues. Born out of the debate has been the development of methods to improve neuronal transplants using less fetal tissue. How long these fetal transplants can be sustained is still questionable. Excitement is also building over another grafting technique which could eliminate dependence on fetuses altogether. Ideally, the best therapy would be to rescue the patient's own dopaminergic neurons and to prevent their death. Testing has begun on the glial-derived neurotrophic factor (GDNF), a protein that could prevent the death of dopaminergic neurons. Should GDNF work in human patients, it could very well make brain-cell grafts passé.*

Researchers Broaden the Attack on Parkinson's Disease

Marcia Barinaga

Science, January 27, 1995

Because there is no cure for Parkinson's disease and no therapy that is effective in the long term, researchers are urgently seeking better treatments. Among the treatments now being investigated, the most controversial seeks to replace the brain neurons that die in Parkinson's disease with transplants of fetal tissue. The treatment shows promise, but it is fraught with political and ethical problems because of its dependence on the use of aborted human fetuses.

But four papers published this month suggest that the search for new treatments is broadening out. Two of the papers suggest ways to improve the neuron transplants or make them independent of the availability of fresh fetuses; the other two take an entirely different tack, suggesting that it might be possible to prevent the neurons at risk in Parkinson's from dying by injecting a growth factor into patients' brains. "It is difficult at this point to guess which [of these therapeutic strategies] will eventu-

ally turn out to be the most efficient," say neural graft pioneer Anders Björklund of the University of Lund, Sweden. "But one should expect that there will be very interesting developments along these lines in the next few years."

Researchers have been highly motivated to find such new strategies because, despite the large amount of effort put into neuronal transplants during the past 8 years, the therapy has suffered from two major handicaps. Because of the ethical controversy, federal funding of the research was prohibited from 1988 until 1992, when President Clinton lifted the ban. During that period, the research that was done (in other countries or with private funding) gave variable results, with some patients showing tantalizing improvement while other continued to decline.

Many researchers think low survival of the transplanted tissue accounts for at least some of the variability. "Right now what we do is put in as much

Reprinted with permission from *Science*, Vol. 267, January 27 , 1995, pp. 455–456.

© 1995 by The American Association for the Advancement of Science.

tissue as possible, and hope and pray that enough cells survive," says University of Chicago transplant researcher John Sladek. The result is that investigators use brain tissue from four to 10 fetuses for each graft.

Better graft survival might reduce the need for so many fetuses, and in a step toward that end, a research team at the University California (UC), San Diego, led by neuroscientist Fred Gage, recently demonstrated a means of boosting the survival of neurons transplanted into the brains of rats. The group mixed the neurons to be transplanted with nonneuronal cells engineered to produce basic fibroblast growth factor (bFGF), a protein that nurtures many types of neuronal cells in tissue culture. They performed their experiment in rats that had been given a Parkinson's-like condition by injecting 6-hydroxydopamine, a neurotoxin that specifically kills the same set of dopamine-producing neurons that die in Parkinson's disease, into their brains.

When the researchers transplanted the engineered fibroblasts along with fetal dopamine-producing neurons (known as dopaminergic neurons) into the animal's brains, they saw a marked improvement in the survival of the transplanted neurons. Ten times as many transplanted neurons survived in the animals that received the bFGF-producing cells as survived in control animals. What's more, Gage says, the Parkinsonian symptoms were completely reversed in the treatment group, while those of the control rats remained essentially unchanged. Those results, published in the January issue of *Nature Medicine*, represent "the most substantial improvements in cell survival" yet, says Björklund.

It's still too early to tell whether bFGF will end up in clinical use, but researchers are optimistic that it or one of the other growth factors that they plan to test will help sustain the fetal transplants. The growth factors "should have clinical relevance, in the near term, in terms of getting better [graft] survival," says Curt Freed, director of the neurotransplant program for Parkinson's disease at the University of Colorado Medical Center.

Even better, says Björklund, would be finding a way of obtaining dopaminergic neurons that is "largely independent of a continuous supply of fetal material." An that's where the second paper, from a group led by Arnon Rosenthal at Genentech, comes in. Rosenthal and his colleague Mary Hynes were searching for the signals that normally trigger differentiation of dopaminergic neurons during development. They teamed up with Marc Tessier-Lavigne of UC San Francisco, who studies a part of the developing nervous system called the floor plate. "We realized the [dopaminergic] neurons were born near the [floor plate]," says Hynes, and so they wondered whether something in the floor plate might trigger their differentiation.

Their hunch proved correct. In the January 13 issue of *Cell*, they report that the floor-plate cells from embryonic rats can induce undifferentiated neuronal precursor cells taken from the rats' midbrain region to become dopaminergic neurons, in both living embryos and the test tube. Even precursor cells that would normally go on to a different fate could be coaxed into becoming dopaminergic neurons instead.

This finding suggests that researchers may be able to grow dopaminergic neurons from undifferentiated precursors in the test tube—which would reduce or even eliminate dependence on fetuses. Working toward that end, Rosenthal says the team is trying to identify the chemical signal in the floor-plate cells that triggers differentiation and is also trying to find a way to grow the precursor cells in continuous culture so they won't have to be collected from fetuses, as they were for the current experiments. The Genentech approach "has the potential to product almost pure dopaminergic neuroblasts [nerve cells' precursors] for grafting," says Chicago's Sladek.

Those are encouraging developments in grafting. But many of those gazing into the future of Parkinson's therapy see a time when there will be therapies that rescue a patient's own dopaminergic neurons, making neuron grafts unnecessary. One candidate as a substance for preventing the death of dopaminergic neurons is glial-derived neurotrophic factor (GDNF), a protein isolated in 1993 by Frank Collins and his colleagues at Synergen Inc., a biotechnology company in Boulder, Colorado. In work reported in two papers in this week's issue of *Nature*,

Lars Olson of the Karolinska Institute in Stockolm, Barry Hoffer of the University of Colorado and their colleagues, and Franz Hefti and Klaus Beck of Genentech and their co-workers show that direct injections of GDNF into the brains of rats and mice can save dopaminergic neurons that have been damaged by either the neurotoxin MPTP or surgical injury.

In the Genentech group's experiments on surgically injured rats, the control animals lost 50% of their dopaminergic neurons, while animals that were treated with GDNF after the injury retained nearly normal numbers of healthy cells. Hoffer's group had similar results with mice treated with MPTP, and in addition found at the GDNF injections also alleviated the Parkinson's-like symptoms induced by MPTP. (The neuron-severing injury used by Hefti's group doesn't produce symptoms in the rats, so they couldn't rest for recovery.)

Although preliminary, these experiments have raised hopes that GDNF will have similar neuron-saving effects in humans. But Freed cautions that it's not yet clear whether results can be extrapolated to humans because the existing animal models are only approximations of Parkinson's. "We simply don't know," he says, what effects growth factors such as GDNF might have in human patients.

We might know soon, though. Amgen Pharmaceuticals bought Synergen last month and plans to move toward clinical trials, although Collins, now at Amgen, says there is no schedule yet for the trials. The first trials will probably involve pumping GDNF directly into the patients' brains, because it is a protein that does not cross the blood-brain barrier. If GDNF proves successful, Amgen and other companies will no doubt search for a small molecule that mimics the protein's effects but can cross the blood-brain barrier and therefore can be administered systemically.

Will this therapeutic approach eventually make brain-cell grafts obsolete? "It is too early to call it," says Hefti, because the approaches to both transplantation and rescuing the patients' own neurons are in such early experimental stages. "Both approaches are very valuable at this point," he says, "and they should be pursued." And it's possible that one of them will prove to be a much better substitute for the existing therapies and their merely temporary respite from the devastation of the disease. ❏

Questions

1. What accounts for the variability of fetal transplants?

2. What would be the advantage of growing dopaminergic neurons from undifferentiated precursors in continuous culture?

3. Why does GDNF have to be injected into the brain?

Answers are at the back of the book.

14

There are an estimated 1.5 million American who are afflicted with Parkinson's disease. A slow path to destruction begins with a tiny section of the brain responsible for supplying the neurotransmitter dopamine to the center of the brain which controls movement. Without dopamine, the body's movements slow down to a snail's pace requiring the patient to become confined to a wheelchair. For some people with Parkinson's, their only treatment is high experimental and controversial technique— a fetal tissue transplant. The procedure is so experimental that its effectiveness has never been fully documented. The source of the tissue, brain cells from aborted fetuses, caused such political fury that the government banned testing the technique in humans during Reagan and Bush presidencies. To overcome these obstacles, researchers are developing alternative sources of tissue, like the patient's own skin and muscle cells, genetically engineered to produce the necessary compounds the brain needs to make dopamine. Until progress is made, fetal tissue transplants remain the only hope to patients with Parkinson's as the scientific and political battles rage on.

Fetal Attraction

Jeff Goldberg

Discover, July 1995

In theory, brain cells that have been killed by Parkinson's disease can be replaced with cells from the brains of aborted fetuses. Now that the necessary politics and the technology are in place, neurosurgeons are about to find out if that theory is correct.

Over the past 11 years, as her Parkinson's disease has progressed, 68-year old Thelma Davis has come to feel trapped in a body that will not move. The symptoms were unalarming at first. Davis told herself that the slight limp in her left leg was nothing serious, that the weakness in her left arm was just her imagination. But although the early signs of Parkinson's disease are so subtle that they are often ignored by patients and misdiagnosed by doctors,

the disease takes a relentless course. Uncontrollable tremors began to appear in Davis's hands and legs. Because the disease affects movements of the jaw and mouth, speech became difficult for her. Little by little, her gait slowed to a flat-footed shuffle, and her face froze into the unblinking, unsmiling mask characteristic of Parkinson's sufferers.

The cause of the disease, which afflicts an estimated 1.5 million Americans, remains unknown, and there is no cure. Scientists do know that Parkinson's casts its imprisoning spell by slowly destroying a tiny section of the brain, the size and shape of a quarter, called the substantia nigra. The substantia nigra supplies the neurotransmitter dopamine to a larger area in the center of the brain, the striatum, which controls movements. As dopamine supplies from the substantia nigra to the stratium dry up, movements slow, become erratic, and finally grind to a halt.

Parkinson's patients, like the rest of us, have plenty of dopamine elsewhere in their bodies: the conundrum is how to get it into their brains. Dopamine can't pass through the blood-brain barrier, a membrane that guards the interior of the capillaries in the brain. Fortunately, the drug levodopa, commonly known as L-dopa, can pass through the barrier, and once it reaches the substantia nigra, cells there convert it into dopamine.

Thanks to L-dopa, Davis was able to lead a relatively normal life for several years, continuing to work as the chief financial officer of a Long Island, New York, mortgage bank. But L-dopa inevitably fails as Parkinson's destroys substantia nigra cells, eventually leaving too few to convert the drug to the neurotransmitter. When her symptoms worsened, Davis reluctantly retired from her job. Now, despite a three-times-a-day regimen of short-acting and timed-release forms of L-dopa, she finds simple tasks like combing her hair and dressing difficult obstacles. She suffers form episodes of "freezing" when the drugs wear off, alternating with spurts of convulsive herky-jerky movements when they shock her system into overdrive—the classic on-off symptoms of advanced Parkinson's disease. Parkinson's itself is not fatal, but many patients die from injuries suffered in falls. Others end up wheelchair bound, unable to move or speak, or succumb to pneumonia.

With her condition deteriorating, Davis has come to the Neural Transplantation Center for Parkinson's Disease at the University of Colorado in Denver for what could be her last hope of recovery—a fetal tissue transplant. In the operation, transplant team leader Curt Freed and neurosurgeon Robert Breeze will implant brain cells culled from aborted fetuses through a thin needle into Davis's brain. To make sure the dopamine reaches the cells that need it, the fetal tissue is grafted into a striatum, where neurons are alive but deprived of dopamine, rather than the substantia nigra, where they're dying. Davis and her doctors hope that as the grafted cells grow and integrate into her brain, they will pump out enough dopamine to replenish depleted supplies and give her back some of her lost mobility.

The tissue for the transplant has been collected, with the consent of the mothers of the fetuses, from private clinics where abortions are performed. It consists of brain cells from the mesencephalon, an area that develops into the substantia nigra and other midbrain structures, dissected from half a dozen six-to-eight-week-old fetuses. The cells must be collected within a narrow window that opens between six and eight weeks into the gestation of each fetus, just before they have fully differentiated into dopamine-producing neurons. "Brain cells at this age can grow just like seeds," says Freed. "They establish root systems in the form of neural connections," regenerating damaged brain circuits. (If the cells are any older, they break up and die during the transplant process.) Because these fetal cells have not yet developed the antigens that trigger an immune response, they also appear to grow without rejection. Over the past 20 days, the cells have been cultured, screened for bacteria and infectious diseases, and tested for levels of dopamine production. Twice during the last three months, Davis's operation has had to be delayed because the tissue was less than perfect.

Davis begins her day by having a metal band bolted to her skull by four pins embedded in the outer layer of bone. A device that looks like a delicate geodesic dome is attached to the band, preparing Davis for her magnetic resonance imagining (MRI) scan. As a gurney inches her forward and back through a powerful doughnut-shaped electromagnet, a scanner detects radio signals emitted by hydrogen atoms in her brain. These signals are reassembled into three-dimensional images by the machine's computer and projected onto a bank of monitors, which Breeze studies intently. From the images, Breeze calculated the angles and routes of the needles that will insert the fetal tissue implants into target sites while avoiding injuries to arteries and vital brain structures. The domelike device provides a grid of reference points for plotting these routes.

After Davis is prepped for surgery and wheeled into the operating room, the dome is replaced with a stereotaxic frame, an awkward-looking device that resembles a large compass or sundial. The frame is a precision measuring tool. Its outer rim contains an array of small holes that can be adjusted to within a fraction of a millimeter to guide the needles delivering the fetal tissue.

Breeze drills four holes a shade smaller than the diameter of a pencil into Davis's forehead and through her skull. As he carefully inserts the needle, Freed prepares the tissue, which has been transported to the operating room in a blue-and-white cooler. The tiny specks of tissue are suctioned into a syringe designed to extrude the tissue in fine, noodlelike strands. These are loaded into a hollow stainless steel tube called a cannula.

The operation is nearly bloodless and, since brain tissue does not register sensation, almost painless as well. Davis is sedated with local anesthetic but remains fully awake. During stereotaxic procedure (which are most often used to obtain biopsy specimens of suspected brain tumors), it's better to keep patients awake and talking, Breeze believes, to help guard against even the remote chance that the needles and catheters inserted in the brain could cause bleeding and precipitate a stroke. While general anesthesia would routinely be used for open brain surgery, when Breeze can see what he's doing, during stereotaxic surgery he works blind, directing instruments into the brain based on computer calculations alone. If the patient were asleep and her brain began to bleed, by the time the doctors notice, it could be to late. So the anesthesiologist keeps up a steady conversation with Davis throughout the operation, carefully listening for any confusion or slurring of speech.

The hollow needle is equipped with an inner stylet, to make it a solid probe that will not cut a core from her brain as Breeze taps the device gently forward. When the needle is in place, Breeze removes the stylet and replaces it with one of the cannulas that Freed has filled with fetal cells, and the infusion begins.

Two hours later, with two small bandages covering the incisions in her forehead. Davis is wheeled out of the operating room into intensive care. She'll go home after four days, but it will be months more before she knows whether the transplant has worked. Fetal tissue transplantation for Parkinson's disease remains highly experimental, and Freed cannot promise a positive outcome.

• • •

Thelma Davis is only the twenty-second patient to undergo the procedure in Denver. Fetal tissue transplants for Parkinson's disease are also offered on a limited basis at Yale, the University of South Florida in Tampa, and the Good Samaritan Hospital in Los Angeles, as well as in England, France, and Sweden, where some of the first experiments with the procedure were performed in the mid-1980s. Roughly 200 such operations have taken place worldwide since the technique was introduced amid a storm of political controversy over the source of the transplanted tissue—electively aborted fetuses. In this country, despite the judgment of a National Institutes of Health advisory committee that fetal tissue transplantation was ethical and promising, government funding to test the technique in humans was banned during the Reagan and Bush presidencies. As a result, clinical studies were limited to a handful of patients. That ban was lifted by executive order during the first days of the Clinton Administration, but now a new debate has surfaced, this time among scientists, over how well the transplants work.

"The moratorium distorted the scientific discussion," says D. Eugene Redmond, the leader of a transplant team at Yale. "To muster the political power to overturn it, the actual scientific accomplishments were somewhat exaggerated."

"There was a presumption that it would work if the ban wasn't there," adds William Freed (no relation to Curt Freed), an NIH researcher. "People thought, 'Well it's banned; it must be something really great.'"

Some patients have shown marked improvement on standard movement tests, such as touching a thumb and forefinger together or tapping their feed, and have resumed many daily activities that most people take for granted: tying their shoes or their ties, vacuuming or driving. They are able to reduce their medication by an average of 50 percent. One of Curt Freed's patients, a California telephone lineman who had nearly lost his ability to speak and was embarrased to eat with friends because he could no longer feed himself properly, celebrated the one-year anniversary of his transplant with a Thanksgiving dinner for 12. Another now enjoys cross-county skiing. "About a third of the patients have had their lives revolutionized," says Freed. "The problem is, the effects are variable."

One in three patients shows only moderate gains, and another third experience no long-lasting benefit at all from the operation. A few even get worse. There are other risks as well. The chance of something going catastrophically wrong during a transplant procedure is small—less than 1 percent that a needle will inadvertently strike an artery or a vital brain area. Nevertheless, in January 1994, Freed's seventeenth patient, a 55-year-old man with an eight-year history of Parkinson's, suffered a stroke in the operating room and died one month later—the first procedure-related fatality.

"We knew this was an odds game," says Freed. "Passing needles into the brain carries risk, and the risk of stroke is about 1 in 500 needle passes. At the time, we'd been doing 14 to 16 needle passes on each patient. With each operation, there was about a 3 percent chance of stroke."

Davis and other patients, many of whom have paid for the $40,000 operation privately, are willing to take that chance. "No other form of treatment holds as much hope as this," says Davis. Neurologists who routinely care for Parkinson's patients remain cautious about recommending the procedure without more proof, however, "Parkinson's disease is slowness of movement, not paralysis," points out Stanley Fahn, a Parkinson's specialist at Columbia Presbyterian Medical Center in New York. "Sometimes with enough excitement or stimulation, sudden movement can return. But this is only transient." Patients in advanced stages, who freeze when they try to cross doorways or are unable to walk across a room without holding onto the walls and furniture, can often negotiate a flight of stairs with ease, ride a bicycle, or catch a ball.

Fahn worries that some improvements observed in transplant patients may be the result of the excitement of undergoing the operation, the mystique of the transplant procedure, and the expectation of getting well. Parkinson's patients may be particularly prone to such placebo effects. Studies of new drugs have shown that as many as 30 percent of Parkinson's patints improve with placebo medications, albeit only briefly. Similarly, neurosurgical answers for Parkinson's disease are also suspect.

In one such surgical technique, called pallidotomy, surgeons destroy a minuscule area in the movement center of the brain—the internal globus palidus—which is located at the base of the brain, just above the spinal column. The procedure was recently reported on *PrimeTime Live* to reduce tremor dramatically in Parkinson's patients. But the *New York Times* followed up with a story detailing how the positive effects many not last, while the operation often leaves patients worse off than before. Skeptics also points out that tremor is only one among many symptoms of Parkinson's. Before L-dopa became a standard treatment, lesion therapies, in which surgeons destroyed parts of various brain structures (usually the thalamus), were also observed to relieve Parkinson's tremors, but these operations didn't relieve slowness of movement in any lasting way.

Autologous transplants, using dopamine-producing cells from a patient's own adrenal glands, also proved to be a disappointment. In the late 1980s hundred of adrenal transplants were performed throughout the world (including about 100 in the United States) after Mexican neurosurgeon Ignacio Madrazo reported startling successes with the procedure. About 40 percent of the patients did experience some initial positive effects, but the benefits of the operation generally vanished before a year had passed. At least part of the problem, according to Freed, was that these cells produce mostly epinephrine and norepinephrine, with only a little dopamine. "They're just not the right kind of cell," he says. "Also, they don't survive well in the brain, because they don't belong there. The tissue around them doesn't provide a supportive environment."

The results suggest the patients were experiencing a placebo effect, but the side effects of the open brain surgery were real enough: respiratory problems, pneumonia, urinary tract infections, sleeplessness, confusion, and hallucinations. Patients had also suffered strokes and heart attacks while undergoing the surgery; about one in ten patients died.

• • •

To put the effectiveness of the fetal tissue operation to the test, Freed and Fahn have joined forces on an unprecedented and controversial study. Forty patients, screened by Fahn at his clinic in New York will undergo transplants in Denver. But to assure that whatever improvements the patients enjoy can legitimately by attributed to the procedure, the study will follow a double-blind, randomized design, similar to a drug trial, in which half the patients will receive a sham operation. Researchers will give the control patients an MRI scan, prep them for surgery, fit them with stereotaxic frames, and drill holes in their skulls; then Breeze will fake the rest of the procedure. Neither the patients nor Fahn, who will be evaluating their progress, will know who has actually received an implant.

"The pacing and atmosphere will be nearly identical to the true tissue implant," says Freed. "The strategy is to do things exactly the same way, maybe even have some tissue set up in a dish so there's time involved in picking up the tissue. We'll drill holes in the skull, the needles will be inserted into the stereotaxic apparatus, all the calculations will be done, the timing will be exactly the same, but the needles will not drop the last 5 centimeters into the brain."

After the operation, Fahn will evaluate the patients by methods ranging from rating their performance on tasks like getting out of a chair to computerized analysis of their videotaped movements. Researchers will also perform positron-emission tomography (PET) scans to evaluate how well the tissue has survived how well the tissue has survived and grown in the patients.

Members of the control group will be eligible to receive real transplants later—providing the procedure passes the test. "We've promised them the treatment. But if it's a bust, they're better off having the sham surgery rather than the real operation," says Fahn.

The four-year, $4.8 million trial is being funded by the National Institutes of Health. This is the first grant awarded for a study of fetal tissue transplantation since the research moratorium ended. A second $4.8 million NIH grant has recently been approved for a similar controlled study of 36 patients who will undergo transplants at the University of South Florida in Tampa.

Freed hopes the studies will provide an unbiased estimate of other value of the procedure. The sham operation presents no additional risk, he says. As an added precaution, the NIH has assigned a Data Safety Monitoring Committee to oversee the studies. The committee has the option of stopping the studies if there are signs of any unexpected complications.

"What you don't want to do, especially with something as dramatic and publicized as a fetal tissue transplantation, is put yourself in the position where you're not sure that what you're seeing is real," says C. Warren Olanow, a neurologist at Mount Sinai hospital in New York. Olanow heads the consortium of investigators that will conduct the second NIH-funded trial. "Is it better to expose a small number of patients to placebo than to forgo a control group and potentially expose hundreds of thousands to a procedure that may not work?"

However, other researchers consider the double-blind studies dangerous and therefore unethical. "There's one in a hundred chance that performing craniotomies on the surgical controls could result in the formation of blood clots. If one of those patients dies, it could set the field back several years," argues neuroscientist John Sladek of the Chicago Medical School.

Controlled trials of fetal tissue transplantation will remain premature, at best, critics believe, until gaping methodological differences between the transplant teams are resolved. The teams disagree on such crucial details as in which of the two sections of the striatum—the putamen or the caudate nucleus—to implant the tissue. The putamen appears to be responsible for a wider range of Parkinson's disease deficits, such as freezing and the inability to walk, but the caudate may govern a number of subtle functions, including eye movement. Researchers also disagree about how much tissue is needed (one to nine fetuses), how to prepare it for transplant (in suspensions, cultures, or cryogenically frozen and thawed specimens), and whether to place large quantities in a few locations or skewer the brain with 15 to 20 micrografts. Even the key question of whether patients should be treated with the immunosuppressant drugs used in organ transplant operations remains undecided.

Improved tissue processing and implantation techniques could also resolve another major concern of researchers, the poor survival of fetal tissue. Recently, when one of Olanow's patients died (of unrelated causes) a year and a half after his surgery, an examination of his brain revealed that hundreds of thousands of fetal cells had survived and formed connections with surrounding brain tissue. But a handful of other autopsy reports and numerous animal studies, showing that up to 95 percent of the transplanted cells die, reinforce the need for further progress. "Although even a few surviving transplanted cells may be enough to produce clinical effects, poor survival may account for variability in the results we've seen so far," says Freed. "People doing kidney transplants were able to get better results simply by improving their surgical techniques and their handling of the organ."

Another problem is the needle-in-a-haystack task of gleaning usable dopamine-producing cells from 2-centimeter-long fetuses often smashed beyond recognition during abortions. The difficulty, Freed once told a Senate subcommittee, is so great that it should be an adequate safeguard against any potential abuses of fetal tissue transplants. Problems with tissue availability may also continue to limit the number of fetal tissue transplants that can be done in Denver and other centers.

• • •

To overcome these obstacles, researchers elsewhere are exploring a number of new technologies, including new versions of autologous transplants. This time the cells used would be the patient's own skin and muscle cells, genetically engineered to produce tyrosine hydroxylase, a chemical the brain uses to make dopamine, as an alternative source of tissue. "From a skin biopsy the size of a quarter we can produce as much tissue in two weeks as you could

harvest from a hundred fetuses," says Krys Bankiewicz, who is working with the California biotechnology company Somatrix Therapy to perfect methods of mass-producing cells for transplant.

In April, doctors at the Lahey Hitchcock Clinic in Burlington, Massachusetts, began trials of cross-species transplants, inserting tissue from five fetal pigs into the brain of a Parkinson's patient. Porcine fetal brain cells are similar to human fetal cells but are more readily available. Experiments are also under way to test "encapsulated" dopamine-producing cells sealed in semipermeable plastic capsules. Because this delivery system is designed to allow dopamine out of the cells while preventing immune destruction, scientists hope unmatched adult tissue or even animal tissue could be transplanted into patients without immune suppression.

But dopamine may not be the whole story, Some researchers believe the fetal transplants may also produce growth factors—chemicals that stimulate nerve cells to sprout. In a recent article in the *Journal of Neurosurgery*, Bankiewicz describes experiments he conducted at the NIH in which he transplanted a variety of fetal cells—none of which produce dopamine, but all rich in growth factors—into rats and monkeys. The result was nearly as good as fetal transplants of mesencephalic tissue, producing "a measurable improvement" in the animals lasting 7 to 12 months.

According to Bankiewicz, the healing powers of fetal tissue grafts stem from a "dual effect" of dopamine and growth factors. Future studies may include trying to boost the effectiveness of fetal issue grafts by infusing additional growth factors or injecting growth factors alone.

For the time being, however, patients seeking a transplant to relieve their Parkinson's disease have only one option—fetal tissue. "The field is still in its early stages. But I'm very optimistic that as our techniques improve we will have a chance of curing Parkinson's," says Freed. "The patients are desperate, and we have no other means of helping them. If they could wait five years, we could probably do it better. But some can't."

Despite the risks and unknowns, there's no lack of volunteers for the trials. Enrollment has been completed in the 40-patient Denver study, and though the 36 patients for the Tampa trial have not yet been selected, Olanow has had no trouble finding willing subjects for the study's preliminary stages. The Denver patients range in age from their early fifties to 75; Olanow expects the Tampa patients to start as young as 35. In both studies half the patients will be under 60 to see whether the patient's age alters the effectiveness of the procedure.

Olanow warns his patients that they must have realistic expectations. "We need people who are absolutely committed to seeing it through. If a patient is unrealistic and he doesn't have a great result, you won't see him again. We need to be able to follow these patients, especially the bad results, because you've got to know what wen wrong as well as what went right."

For Thelma Davis, as she recovers at her home on Long Island after the long flight back from Denver, waiting to see if the transplant worked is going to be tough. "You try to rationalize the situation, the facts, that it's going to take months and may not be a cure. But it's not logical," she says. "I just hope I have the patience." ❑

Questions

1. What part of the brain is destroyed by Parkinson's disease?

2. What is L-dopa?

3. What are autologous transplants?

Answers are at the back of the book.

Section Three

The Brain, the Nervous System, and the Properties of Neurons that Serve Them Both

15 *Many people suffer with chronic pain that is enigmatic. Such miserable pain can often be debilitating with no known cause. But despite the unknowns, neurobiologists believe that research is beginning to find clues into what might be triggering this purposeless pain. There is growing evidence that nerve growth factors, located at sites where touch nerves are clustering, might be responsible for cross-wiring with sympathetic nerve fibers that carry information about pain to the spinal chord and to the brain. Radical new strategies are already being developed that will block the production of nerve growth factors and dampen the pain sensation for many patients with enigmatic pain.*

Racked with Pain

Elizabeth Pennisi and Rachel Nowak

New Scientist, March 9, 1996

Millions of people suffer from incurable pain that has no immediate cause. Elizabeth Pennisi and Rachel Nowak find the reason may lie with crossed connections in the nervous system.

Imagine the pain of someone holding a burning cigarette to your naked skin. Now imagine the pain with out the cigarette... Imagine a sensation akin to an electric shock shooting through your shoulder blade. Imagine deep, unremitting backache that leaves you permanently debilitated. Finally, imagine a skin sensitivity so acute that even the draft from an open door is painful.

This is the nightmare world of enigmatic pain. Here, the pains that make people's lives misery for decades serve no apparent function. They do not, for example, warn of potential dangers—of fire, say, or a twisted ankle—nor do they ensure that a person slows down and heals an injury. And because drugs are no use, many sufferers end up doing rounds from neurologist to chiropractor to acupuncturist, losing money and gaining little relief. Worse still, because such pain has no obvious and immediate cause, there is always the unspoken charge that the pain is imaginary. For some, suicide offers the only respite.

Nor do doctors escape the anguish. "When a doctor sees the same patient coming back every few months, and there is nothing [he or she] can do to help the pain, both the doctor and the patient get desperate," say Clifford Woolf, a neurobiologist at University College London. Among themselves, doctors classify people suffering from this enigmatic, chronic pain as "heartsink" patients.

Now, an end to at least some of that misery and frustration may be in sight. Neurobiologists believe that certain diseases and injuries (some of which are long forgotten by the time the pain becomes a handicap) trigger crossed wires deep within the nervous system by stimulating the production of chemicals that encourage new growth. Those crossed wires link nerve fibres carrying everyday messages about touch and pressure from the body's surface with the pain pathways to the brain, and they may be at least partially responsible for triggering the bouts of inexplicable, unpredictable pain.

Painful Times

And the recognition that physiology not psychology is to blame is more than a panacea for the battered egos of patients and doctors. It could also one day translate into radical new therapies that tackle enigmatic pain both by attempting to prevent the cross-wiring from occurring in the first place, and undoing the cross-wiring one it has taken place.

Most pain is vital to survival. The sharp pain associated with, say, touching a hot iron, is the body's way of warning that it is under threat. It's called protective pain. Signals pass from pain receptors in the skin, along nerve fibres to the spinal cord, and on the brain where they provoke the sensation of pain and, hopefully, evasive action. A second type of pain, called reparative pain, makes sure that once the damage is done the body gets a chance to heal. It's a sickening, aching sensation that is often accompanied by extreme sensitivity to normal amounts of pressure, heat, or light, and it only disappears after sufficient time has passed to ensure that the healing process has got well under way.

Roughly 2 percent of the population, however, suffer the third, useless sort of pain that neither acts as a warning nor helps with the healing process. And this kind of pain costs millions—perhaps billions—worldwide in lost productivity and healthcare. One estimate set a figure of up to 365 million days of pain in the UK alone.

Studying the mechanisms that underlie inexplicable chronic pain is difficult in humans, not least because the condition is notoriously variable, so Woolf and his colleague Richard Coggeshall at the University of Texas Medical Branch at Galveston turned to the lab rat. The two researchers cut or crushed the sciatic nerve—a large nerve in the leg that carries information about touch, temperature, and pain, and relays instructions to the muscles—of the anaesthetised animals. They studied how the nerve fibres in the rats' spinal cords changed in response to that type of injury.

In cross section, a darker butterfly-shaped region is imprinted on the eggshaped spinal cord. The butterfly's wings are the spinal cord's ventral and dorsal horns, the latter of which is divided into five microscopic layers. In four of the dorsal horn layers, nerve fibres bringing messages about touch from the skin make contact with the nerve cells that give rise to the massive nerve tracts that relay the messages on to the brain and other parts of the spinal cord. The second layer of the dorsal horn, however, is reserved mainly for incoming nerve fibres relaying pain messages.

Woolf and Coggeshall discovered that within two weeks of the initial damage, a mass of extra nerve fibres has sprouted into the second layer of the rat's dorsal horn. Using an electron microscope to study the connections between fibres, and electrical recordings to find out what sort of nerve fibres had encroached on the second layer, the researchers concluded that there was a massive fifteen-fold increase in the number of touch fibres making connections with the nerve cells that transmit pain information to the brain, and that normally only receive inputs from pain fibres.

"Paradoxically, the system that reports on tissue damage [to the brain], is itself affected by tissue damage," says Stephen McMahon, who studies pain at the St Thomas's campus of the United Medical and Dental Schools in London.

Woolf thinks it is likely that the same thing happens in patients who have had a nerve damaged in an accident or by diseases such as diabetes. "There's a rewiring so that whenever the fibre is activated by touch or vibration, the central nerve cells in the spinal cord think they are getting an input of pain or noxious damage, [and] the person experiences pain," he says.

Elspeth McLachlan, a pain researcher at the Prince of Wales Medical Research Institute in Sydney, agrees with Woolf, and points out that the rats' injuries roughly mimic the sort of damage that often leads to chronic unexplained pain is a person. Cutting the sciatic nerve is equivalent to "going through the window in a car crash, and having a great pane of glass cut through you leg," she says, and it is just "such injuries that most commonly precede chronic intractable pain in people."

The "rewiring" theory also offers clues about why certain surgical procedures that until recently were fashionable options for people with chronic pain can actually make the conditions worse. "It suggests that if you cut a peripheral nerve to relieve pain, you could get central rewiring that actually causes more pain," say Coggeshall.

Nerve Sprouts

Unexpectedly, the rat results may contain a pleasant surprise for humans: within nine months, the deviant nerve connections had disappeared. "It was a tremendous surprise. We had thought that once the whole system had rewired that was it. But this means it is reversible," say Woolf. Indeed, in patients, chronic pain can suddenly disappear of its own accord, suggesting that what happened in the rats may also happen in humans. But in many cases, chronic, untreatable pain plagues patients for years on end. For that reason, Woolf and Coggeshall, together with scientists at Regeneron, a biotechnology company in Tarrytown in New York state, are now searching for the chemicals that stimulated the nerve fibres to sprout in the first place because blocking those chemicals could prevent the development of chronic, purposeless pain in the first place. They are also looking for any chemicals that might have helped break the aberrant connections in case they are able to provide a way of alleviating pain once it has taken hold.

There is already good evidence that a nerve "fertiliser" with the prosaic name of nerve growth factor, or NGF, triggers a second type of pain-related cross-wiring in which sympathetic nerves, whose task is to regulate organs like the heart, blood vessels, and stomach, connect to the types of nerve fibres that carry touch information to the spinal cord.

McLachlan and Wilfred Jänig at the University of Kiel in Germany looked at the sympathetic nerves of rats with cut sciatic nerves. They found that in damaged animals, the sympathetic nerves that usually just control the blood vessels send out shoots that encircle the cell bodies of the touch nerves clustering in knots called ganglia just outside the spinal column. In McLachlan's rats, not only were the two types of nerves connected by the sprouts, but when the sympathetic nerves were stimulated with electricity, the touch fibres responded by becoming either more or less active. Because those touch nerve fibres are the very same ones that Woolf and Coggeshall discovered encroaching on the pathways in the spinal cord that relay pain information to the brain, it seems likely that nerve impulses passing along sympathetic nerve fibres may also trigger enigmatic pain, says McLachlan.

At roughly the same time as McLachlan was making her discovery, Brian Davis and Kathryn Albers at the University of Kentucky in Lexington were genetically engineering mice to produce too much NGF in their skin in an effort to find out how the chemical regulates the development of the nervous system in a growing animal.

The effects of the extra NGF turned out to be dramatic and surprising. Not only are the engineered mice extremely sensitive to the pain of having their feet poked with nylon fibre, their nervous systems are wired to the same strange plan as McLachlan's rats. NGF is secreted by the stumps of damaged nerves, so it is likely that the chemical also triggers the sprouting of the sympathetic nerves in the rats with cut sciatic nerves.

In the case of the engineered mice, says Davis, the ends of the sensory nerve fibres pick up NGF in the skin and transport it to their cell bodies, where presumably it leaks out and stimulated the growth of the nearby sympathetic nerves. "At the moment, we don't know whether the NGF keeps going into the spinal cord," he explains. But if it does, NGF would also be implicated in forging those mysterious connections discovered by Woolf and Coggeshall.

The discovery that pain, nerve damage, nerve fetilisers, and the rewiring of the nervous system are inextricably linked in animals has brought a new impetus to pain research, say McMahon. He is careful to add a rider that animal studies have also linked chronic pain to a range of other abnormalities, including changes in the pain receptors in the skin, in the nerve fibres that transmit the pain messages from the body's surface, and in the chemicals that transmit messages from one nerve cell to the next. "It's not yet clear which is the most significant in terms of pain sensation for the patient," says McMahon.

But despite the unknowns, there is enough evidence about nerve rewiring for researchers and doctors to start devising new strategies for combating this purposeless pain. Traditionally, doctors treat severe pain with opiate-like drugs. However, even although these drugs hamper the transmission of pain messages in the spinal cord, they are no use for people suffering from the wrong sort of pain. Now, they can start hoping for drugs that will either block the production of NGF and other nerve fertilisers immediately following a severe injury at the actual sites where rewiring is likely to occur or that will undo the rewiring once it has happened. "It's a long way down the line," says McLachlan; "but now it is conceivable."

Meanwhile, for those tortured by what once seemed to be entirely inexplicable pains, there is at least the comfort of knowing that it's their nerve cells—not their minds—that have gone haywire. ❏

Questions

1. What are the three types of pain?

2. Why is enigmatic pain difficult to diagnose and treat?

3. What is the "rewiring" theory of enigmatic pain?

Answers are at the back of the book.

16

In biomedicine, it is the decade of the brain. There has been an explosion of discoveries about the brain which are raising new possibilities for identifying, treating and preventing brain disorders that have until now remained a mystery. Disease-producing genes associated with Huntington's, Alzheimer's, and Lou Gehrig's diseases, epilepsy and muscular dystrophy have been identified and scientists are making great strides in elucidating the molecular mechanisms behind their progression. New drug therapies may be developed for their neuroprotective effects which can guard brain cells from premature death or even help them regenerate. Before the end of the decade, researchers will have gathered enough new findings to strengthen the very foundations in the study of the human brain.

Revealing the Brain's Secrets

Kathleen Cahill Allison

Harvard Health Letter, Special Supplement, January 1996

Is space truly the final frontier? Not according to scientists who are probing what they call the most complex and challenging structure ever studied: the human brain. "It is the great unexplored frontier of the medical sciences," said neurobiologist John E. Dowling, professor of natural science at Harvard University. Just as space exploration dominated science in the 1960s and 1970s, the human brain is taking center stage in the 1990s.

It may seem odd to compare an organ that weighs only about three pounds to the immensity of the universe. Yet the human brain is as awe-inspiring as the night sky. Its complex array of interconnecting nerve cells chatter incessantly among themselves in languages both chemical and electrical. None of the organ's magical mysteries has been easy to unravel. Until recently, the brain was regarded as a black box whose secrets were frustratingly secure from reach.

Now, an explosion of discoveries in genetics and molecular biology, combined with dramatic new imaging technologies, have pried open the lid and allowed scientists to peek inside. The result is a growing understanding of what can go wrong in the brain, which raises new possibilities for identifying, treating, and perhaps ultimately preventing devastating conditions such as Alzheimer's disease or stroke.

"The laboratory bench is closer to the hospital bed than it has ever been," said neurobiologist Gerald Fischbach, chairman of neurobiology at Harvard Medical School, where the brain and its molecular makeup are a primary focus of research.

One important challenge is to understand the healthy brain. By studying brain cells and the genetic material inside them, scientists are discovering how groups of specialized cells interact to produce memory, language, sensory reception, emotion, and other complex phenomena. Figuring out how the healthy brain goes about its business is an essential platform that researchers need in order to comprehend what goes wrong when a neurological disease strikes.

There have also been great strides toward elucidating some of the common brain disorders that rob people of memory, mobility, and the ability to enjoy life. The most promising of these fall into several broad categories.

• The discovery of disease-producing genetic mutations has made it possible not only to diagnose inherited disorders, but in cases such as Huntington's disease, to predict who will develop them. These findings have also pointed the way toward new therapies.

• Insights into the programmed death of nerve cells may lead to drugs that can halt the progession of degenerative diseases or contain stroke damage.

• Naturally occurring chemicals that protect nerve cells from environmental assaults may hold clues about preventing disease or reversing neurologic injury.

• Information about brain chemistry's role in mood and mental health has already helped people burdened by depression, for example, and is expected to benefit others as well.

Genetics Opens a New Door

Discovering a gene associated with a disease is like unlocking a store house of knowledge. Once researchers have such a gene, they may be able to insert it into experimental systems such as cell cultures or laboratory animals. This makes it easier to discern the basic mechanisms of the disorder, which in turn helps scientists figure out what diagnostic tests or therapies might be best. When a new treatment is proposed, genetically engineered models of human diseases make testing quicker and more efficient.

In recent years, scientists have found abnormal genes associated with Huntington's disease (HD), Alzheimer's disease (AD), amyotrophic lateral sclerosis (ALS or Lou Gehrig's disease), one form of epilepsy, Tay-Sachs disease, two types of muscular dystrophy, and several lesser-known neurological conditions.

A decade-long search for the HD gene ended in 1993, when Harvard researchers Marcy MacDonald

and James Gusella, working with scientists at other institutions, identified a sequence of DNA that produces symptoms of the disease if it is repeated enough times. Huntington's is a progressive and ultimately fatal hereditary disorder that affects about 25,000 people in the United States. It typically strikes at midlife, and the researchers discovered that the more copies of the sequence a person inherited, the earlier symptoms show up.

Scientists quickly developed a highly reliable assay that enables people with a family history of HD to find out if they or their unborn fetus harbors the dangerous mutation. But because no cure for the disease exists, few people have rushed to have themselves tested.

Demand might increase, however, if scientists can use the HD gene to design effective treatments. Genes contain the assembly instructions for proteins, the molecules that carry out the day-to-day operations of the body. Scientists strive to identify the protein made by a disease-producing gene and to figure out what it does, which in turn helps them understand the event that initiates the disease process.

The HD gene codes for a protein that appears to contribute to the premature death of certain neurons. It is the loss of these cells that results in the involuntary movements and mental deterioration typical of Huntington's. When researchers know more about this protein, they may be able to develop drugs or other therapies that could slow the onset of symptoms or even block them entirely.

A Downward Spiral

The gradual extinction of certain brain cells is also the underlying cause of Alzheimer's disease. In this case, the impact is progressive loss of memory, changes in personality, loss of impulse control, and deterioration in reasoning power. Under the microscope, the brains of people who died with AD are studded with abnormalities called amyloid plaques and neurofibrillary tangles. About 20% of all AD cases are inherited, and these people develop symptoms earlier in life than those with the more common form, which typically appears well after age 65.

In recent years, scientists have discovered several

different genetic mutations that can cause the unusual, inherited form of AD. One of these abnormal genes has successfully been introduced into mice by researchers at several pharmaceutical companies, and experts believe that this animal model will help them understand how all forms of the disease progress at the cellular and molecular level.

So far, it looks as though some of the animal's brains develop amyloid plaques like the ones that build up in humans. Long-standing doubt about whether plaques cause symptoms may be resolved by future observations of whether these genetically engineered mice show signs of memory loss. If there is a strong correlation between amyloid accumulation and symptom severity, these mice will be used to test drugs that might keep plaques from forming.

The Cell Death Story
Unlike other types of cells, nerve cells (neurons) are meant to last a lifetime because they can't reproduce themselves. Struck by the realization that abnormal cell death is the key factor in neurologic problems ranging from Alzheimer's to stroke, scientists have embarked on a crusade aimed at understanding why nerve cells die and how this might be prevented.

It's normal to lose some brain cells gradually. Trouble arises when large population of cells dies all of a sudden, as in a stroke, or when too many of a certain type die over time, such as in Alzheimer's or Parkinson's (PD) disease. While some scientists remain skeptical that inquiries into cell death will ever lead to effective means form preventing or treating neurodegenerative diseases, many others are enthusiastically pursuing this line or research.

Some scientists are racing to develop *neuroprotective* drugs that could guard brain cells against damage and death or even help them regenerate. There are many different ideas about how to do this.

For example, although Harvard scientists have identified the gene for HD and the protein it makes, they don't understand the mechanisms that lead to symptoms. One theory is that a phenomenon called *excitotoxicity* is responsible, and that Huntington's is only one of many diseases in which this process plays a role.

The idea behind excitotoxicity is that too much of a good thing is bad for cells. Glutamate, for example, is an ordinarily benign chemical messenger that stimulated certain routine cellular activities. Under extraordinary circumstances, however, "cells can be so excited by glutamate that they wear themselves out and die," said John Penney Jr., a neurologist at Massachusetts General Hospital and a Harvard professor neurology.

Sending a Signal
One of the many types of doorways built into the walls of nerve cells is a structure called an NMDA receptor. One of its functions is to allow small amounts of calcium (a substance usually shut out of the cell) to enter it. This happens when the NMDA receptor is stimulated by glutamate. If excess glutamate is present, too much calcium rushes in—an influx that is lethal to the cell.

Someday it may be possible to halt the advance of Huntington's by injecting drugs which block the NMDA receptor so that calcium can't get in. In animal experiments, scientists have demonstrated that such receptor-blocking agents can keep brain cells from dying. Harvard researchers are seeking approval for a clinical trial that will test such neuroprotective drugs in patients with symptomatic disease. If participant obtain any relief from this treatment, the next step will be to determine whether this approach can prevent symptoms in patients who have the gene but do not have symptoms.

Scientists also hope that neuroprotection can be used to limit brain damage due to stroke. When a stroke shuts down the supply of blood to part of the brain, neurons in the immediate area die within minutes. Over the next several hours, more distant cells in the region are killed as excitotoxic signals spread. In an effort to limit the extent of brain damage, researchers are currently treating small numbers of patients with intravenous doses of experimental agents such as NMDA receptor blockers and free radical scavengers. Other neuroprotective agents under development include protease inhibitors, nitric oxide inhibitors, and nerve growth factors.

"Our dream is a safe and effective neuroprotectant that can be given to the stroke patient in the ambulance or shortly after arrival in the emergency room," said neurologist Seth Finklestein, an associate professor at Harvard Medical School who conducts basic research at Massachusetts General Hospital. "That's the holy grail of neuroprotective treatment."

Applications for Alzheimer's

Neuroprotection is also making waves in Alzheimer's research, as scientists strive to inhibit the type of cell death that typifies this disease. One group of investigators has identified several *peptides* (small protein molecules) that block the formation of amyloid plaque in the test tube, said neurobiologist Huntington Potter, an associate professor at Harvard Medical School. The researchers hope to test these peptides in humans.

Brain cells manufacture several neuroprotective chemicals on their own, which scientists call *neurotrophic* or nerve growth factors. These small proteins may hold the key to keeping cells alive even in the face of stroke, degenerative diseases, or even spinal cord injury.

For example, several different neurotrophic factors are being tested in the laboratory to determine if they could protect the dopamine—producing cells that die prematurely in people with Parkinson's disease. Other uses are being studied as well, and some researchers anticipate that these chemicals will be tested in humans before the decade draws to a close.

Mood, Mind, and Brain Chemistry

Scientists have discovered that a surprising number of mental disorders, from depression to schizophrenia, are the result of brain chemistry gone awry. And this understanding has led them to design new medications for treating specific mental disorders and behavior problems.

The best known of this new breed of drugs is fluoxetine (Prozac), one of several selective serotonin reuptake inhibitors (SSRIs). It was possible to design these agents, which are widely prescribed to alleviate depression and related disorders, only after scientists came to understand how nerve cells communicate at the molecular level.

Each nerve cell has an *axon*, a long branch that reaches out and touches other nerve cells. A tiny space called a *synapse* separates the axon terminal (which sends a message) and the cell body (that receives it), and this is where the action is. The sending cell releases *neurotransmitters* (chemical messengers) into the synapse which either excite or inhibit a receiving cell that is equipped with the proper receptors. Messages pass from cell to cell in this manner, eventually leading to a physiologic action. In each synapse, the cell that sent the message sops up leftover neurotransmitters and stores them for future use. People who are depressed have less serotonin than those who aren't, and the SSRIs block the reuptake of this chemical, thereby boosting the effect of small amount on the receiving cell.

But Prozac and its relative are only the tip of the iceberg. As researchers work to understand the roles of different chemical messengers and the highly specific receptors that bind them, a whole new approach to the treatment of mental disorders is evolving. The identification of highly specialized receptors is already paving the way for ever more specific drugs to treat these conditions.

Schizophrenia therapy is a case in point. As devastating as this form of mental illness is, treatments have sometimes appeared worse than the disease. Until very recently, the only drugs that relieved symptoms could also lead to spasmodic, uncontrollable movements known as *tardive dyskinesia*. This is because these agents block all types of receptors for dopamine, a neurotransmitter that is a key player in normal movement as well as in this mental disorder. Now there is a new drug for schizophrenia, clozapine, that blocks only a small subclass of dopamine receptors. It relieves symptoms of the illness in some people without leading to abnormal movements. Still, it can have other serious side effects.

Tailored to Fit

The bottom line for the treatment of behavior and emotional disorders may be that drugs will become ever more specialized. Just as computers now help salespeople fit blue jeans to the individual purchasers, it is not inconceivable that psychopharmacologists may someday tailor drugs to the needs of each patient.

What does the future of brain research hold? Dr. Dowling anticipates that medications that can slow the process of degenerative disease, correct the chemical imbalances that cause mental disorders, prevent stroke damage, and repair spinal cord injuries may all be on the horizon. "We have learned so much about the cellular and molecular aspects of the brain," Dr. Dowling said. "We stand at a time of great opportunity, when we can take tremendous advantage of these things and turn them into practical clinical therapies." ❏

Questions

1. What have scientists discovered about the Huntington's disease gene?

2. Why are neuroprotective drugs being developed?

3. What are neurotrophic growth factors?

Answers are at the back of the book.

17 *Open any physiology textbook and you'll learn that action potentials are propagated in one direction away from the original site of activation. While this idea is true, it is being revised to include the newest findings which suggest that in the brain, dendrites also support backpropagation of the action potential from its initial site. The cell membrane depolarizes and thus elevates intracellular Ca2+ in the dendrite, much like what occurs in synaptic plasticity. Furthermore, these retrograde dendritic potentials could also be playing an important role in synaptic integration and learning.*

Action Potentials in Dendrites: Do They Convey a Message?

Heinz Valtin

News in Physiological Sciences, April 1996

The textbook view of dendritic trees of neurons in the central nervous system states that they are passive structures. Synaptic potentials were said to spread passively along the dendrites and summate at the cell body; if the resulting depolarization reached a critical level, an action potential would then be generated at the initial segment of the axon. The message of the neuron is thus encoded as action potentials that are propagated orthograde along the axon. Until recently, not much thought was given to the fate of the action potentials that simultaneously spread backwards into the dendritic tree. If indeed the dendrites behave as passive cables, such retrograde action potentials should decay very rapidly along the proximal dendrites, because the potentials have a fast time course. Therefore, one would expect little local interaction between the synaptic potentials and what was left of the action potential in the dendritic tree, except in the most proximal compartments of the tree.

This prediction was tested in technically demand-ing experiments in which recordings were obtained simultaneously from the cell body and from its dendrites. These experiments have revealed unequivocally that, in certain areas of the brain, dendrites are indeed excitable and that they actively support backpropagation of the action potential from its site of initiation onto the dendrites. Areas for which this fact has been established include CA1 pyramidal neurons in the hippocampus,[3] pyramidal neurons in the neocortex,[2] and developing motoneurons in cell culture.[1] The dendritic action potentials were, in general, broader than somatic action potentials, and their amplitudes decreased with increasing distance from the cell body. Inasmuch as the dendritic action potentials were blocked by tetrodotoxin (TTX, an inhibitor of Na^+ channels), it was concluded that Na^+ conductance is responsible for these retrograde potentials.

Neither injection of current into dendrites nor excitatory synaptic potentials could elicit action potentials in dendrites. Rather, these potentials were

Reprinted with permission from *News in Physiological Sciences*, Vol. 11, April 1996, p. 101.

always initiated first in the cell body or close to it. All the neurons in which dendritic action potentials have been demonstrated appear to have one distinct site (probably the initial segment of the axon), which is equipped with a high density of Na^+ channels and which can initiate action potentials. Starting from this location, the action potential sets off in two directions, along the axon and back into the dendritic tree.

Although the functional meaning of the orthograde action potential traveling along the axon is undisputed, the meaning of the retrograde dendritic potential is much less obvious. Most likely, the latter conveys information about the level of activity of the neuron to postsynaptic sites, many of which are located on the dendrites. This possibility is supported by the observation that the backpropagated potential leads to a substantial elevation of intracellular Ca^{2+}, probably through voltage-activated Ca^{2+} channels.[1-3] Since the backpropagation of action potentials varies in different dendrites of the same neuron[1] or even may stop at dendritic branch points,[3] each branch of a dendritic tree may encounter very different voltage and Ca^{2+} signals during repetitive firing of the neuron. Depolarization and elevated Ca^{2+} in the dendrites may be involved in the induction of synaptic plasticity, and pairing of pre- and postsynaptic activity may lead to so-called Hebbian learning, the phenomenon in which synapses are strengthened when both the pre- and postsynaptic neurons are activated simultaneously. Thus the retrograde dendritic potentials appear to convey a message that may be most relevant for synaptic integration, learning, and plasticity during development of the organism.

These new findings suggest that a single neuron with its extensive dendritic tree has the capability of computing more complex functions of its input than simply "integrating and firing," and that therefore most textbooks need revision.

References

1. Larkum, M.F. , M. Rioult, and H.-R. Luscher. Propagation of action potentials in the dendrites of neurons from rat spinal cord slice cultures. *J. Neurophysiol.* **75**: 154–170, 1996.
2. Markram, H., P.J. Helm, and B. Sakmann. Dendritic calcium transients evoked by single back-propagating action potentials in rat neocortical pyramidal neurons. *J. Physiol. Lond.* **485**: 1–20, 1995.
3. Spruston, N., Y. Schiller, G. Stuart, and B. Sakmann. Activity-dependent action potential invasion and calcium influx into hippocampal CA1 dendrites. *Science Wash. DC* **268**: 297–300, 1995. ❏

Questions

1. What are dendritic action potentials?

2. How different are dendritic action potentials from somatic action potentials?

3. What purpose do retrograde action potentials serve in the neuron?

Answers are at the back of the book.

18 *While studying the brain, scientists have discovered that it has the potential to repair its own tiny injuries. To be completely effective as therapy for a brain damaged by an injury or a disease such as Hungtington's or Parkinson's, new developments must be made to promote cell division. Stem cells already present in the brain are being used to perform a type of gene therapy to correct a cell deficiency that exists in a brain with Tay-Sachs. Another approach involves the use of epidermal growth factor to stimulate cells to differentiate into other brain-cell types like neurons. Both strategies could eventually assist the brain in healing itself by replacing the specialized tissue that has been damaged by a stroke or serious disease.*

Brain, Heal Thyself

Josie Glausiusz

Discover, August 1996

Once brain cells die—from aging, injury, or disease—they can't be replaced. Or can they? Researchers are trying to stimulate new growth in old brains.

Deprived of our blood-forming stem cells, we would all quickly die. These bone-marrow cells replenish red and white blood cells day in and day out for decades. The skin, liver, gut, and perhaps other organs are also thought to have their own stem cells that replace injured and dead cells. Not so the brain: The conventional wisdom has long been that it doesn't have stem cells—perhaps in part because it would have a hard time holding on to memories if its cells were constantly being replaced. Instead the brain starts out with more cells than it ordinarily needs in a lifetime. "Nature gives you too many brain cells to start with and assumes that you won't do anything silly like get into a boxing ring or ride a motorcycle without a helmet," says Samuel Weiss, a neuroscientist at the University of Calgary in Canada.

"And in most cases nature has done well, because most of us don't need replacement."

Nevertheless, the conventional wisdom on brain stem cells is changing these days. Although no one has yet conclusively isolated stem cells from an adult mammal's brain, Weiss and other researchers have induced mouse brain cells to act like stem cells in the lab. And they have found good reason to hope that it may one day be possible to get cells in the adult human brain to act like stem cells—and perhaps replace tissue that has been damaged by stroke or by a disease such as Huntington's or Parkinson's.

One of the leaders in this new field is Evan Snyder, at Harvard Medical School. In 1992 he announced that he and his colleagues had removed "stemlike" cells from the brains of newborn mice. Specifically, the cells came from the cerebellum—a motor-coordinating area of the brain that continues developing for a brief postnatal period. These immature cells were amorphous and flat, lacking the long, delicate connecting fibers—the axon and dendrites—of mature neurons. Under normal circumstances these cells would rapidly differentiate into specialized cells and would no longer reproduce themselves. But Snyder infected them with a retrovirus carrying a gene that prompted the cells to divide. Not only did the cells reproduce, they also began spinning off the three main types of mature brain cells: the message-carrying neurons; astrocytes, cells that surround the capillaries, forming the blood-brain barrier; and oligodendrocytes, which make the myelin that insulates neurons.

Although their genesis was somewhat artificial, Snyder claims that his manipulated cells meet the requirements of true stem cells: they can reproduce and maintain themselves, and they can give rise to all the major cell types in the brain. But were they just a laboratory curiosity? To find out, Snyder injected the genetically engineered cells into the brains of newborn mice, with a genetic marker that allowed him to track them. (The marked cells turned blue when exposed to a special stain.) After the mice matured, he killed them and examined their brains.

Snyder found that the marked cells had indeed differentiated into neurons and other brain cells—their destiny dependent on the site at which they had settled—and some had formed normal synaptic connections with existing brain cells. What's more, after differentiating, the cells had ceased dividing, just as normal brain cells would—possibly because of some innate brain signal that dampens division. To date, Snyder has injected his stemlike cells into more than 1,000 mice without once seeing the uncontrolled cell growth that makes a tumor.

Snyder's long-term goal, however, was to see whether his implanted cells could repair some kinds of brain damage. And in recent experiments, he has found that they probably can. For example, when he injected the cells into newborn mice with artificially induced stroke, the cells migrated into damaged areas. Some differentiated into neurons and oligodendrocytes, the cells most commonly injured when the oxygen supply is cut off, as it is in a stroke. Snyder thinks that the cells may migrate and mature so readily because they are responding to developmental signals analogous to those that occur in the embryo—growth factors, perhaps, that in this case

are put out by dying neurons or their neighbors. Ordinary mature brain cells, he speculates, have lost the ability to respond to such signals, or the signals may somehow be suppressed.

In his latest research, Snyder and his colleagues are using his "stem cells" to perform a type of gene therapy. They spliced into the cells a gene that codes for an enzyme missing in children with Tay-Sachs disease. This enzyme breaks down a cellular waste product in the brain. Without the enzyme, the waste accumulates in the brains of children with the disease, causing severe mental retardation and death. Snyder found that once inserted into mouse brains, the genetically engineered cells began producing the enzyme at levels thought to be sufficient to alleviate symptoms of the disease in humans. In a brain with Tay-Sachs, he thinks, the stem cells might naturally tend to spread and produce their crucial enzyme throughout the damaged brain.

Weiss, meanwhile, has taken a different approach to cell repair in the brain. He has been working with cells taken from the subependymal layer, at the core of the brain. In mice, this region produces specialized cells that replace worn-out cells in the olfactory bulb, the part of the brain that controls the sense of smell. Weiss has found that by treating subependymal cells with a protein called epidermal growth factor, or EGF, the cells, like those in Snyder's experiments, reproduced both themselves and the three major brain-cell types. Weiss says that both his and Snyder's approaches promote cell division, his method by an external signal from EGF, and Snyder's from an internal genetic command. More research, he says, will determine which is the more effective strategy. Both, however, take advantage of the fact that actively dividing cells have not yet differentiated into specialized tissue.

Recently, Weiss and his colleagues Constance Craig and Derek van der Kooy of the University of Toronto have found that injection of EGF into mouse brains spurred the growth of new neurons. These cells spread into regions near the subependymal layer, including the striatum, which is involved in regulating motor functions. This is significant, because in people with Huntington's disease, neurons in this region die. "Something that

I would consider to be very primitive—simply infusing EGF—seems to have the potential to replace the neurons that are lost in Huntington's disease," says Weiss.

For now, the gap between experiments with laboratory mice and human cell therapy for brain damage is enormous. Snyder and Weiss both believe, however, that their experiments show that the human brain has the potential to repair itself, and that it may indeed even have its own stem cells, only in numbers too small to be effective for anything but the repair of tiny injuries. Infusing it with EGF might be one way to help it; transplanting cells that have been taken from the brains of human accident victims, and that have been manipulated to become stemlike, might be another.

"Sometimes, when the brain is really massively damaged," says Snyder, "it tries to evoke these same mechanisms but just can't quite do it to the extent that you care about. What I take away from this is that the brain wants to repair itself—there are cries for help, so to speak. Now, if we understand the language of those cries, I think we can jump into that breach and help out, either by supplying more of the factors that the brain is making at a low level or additional stem cells to augment the brain's own supply." ❑

Questions

1. What is the job of stem cells in the body?

2. Are there stem cells in the brain?

3. How could stem cells be used to treat the brain with Tay-Sachs?

Answers are at the back of the book.

19

The hormone, leptin, is sending scientists back to the bench scratching their heads and wondering what its real function is. This is due in part to the fact that the theories originally developed to explain leptin's role in the body have remained unproved. Produced and released by adipose cells, it was believed that leptin works as a signaling factor involved in the control of body weight. One major finding to disprove this theory was that in genetically normal animals, leptin does not reduce weight and food intake by acting as a brain receptor for controlling eating behavior. While many questions still remain unanswered, scientists need to learn more about the leptin protein whose role thus far remains illusive.

What Is Leptin for, and Does It Act on the Brain?

Richard J. Wurtman

Nature Medicine, May 1996

Only hermits are unaware, by this time, that adipose cells secrete a hormone, leptin;[1] that its secretion and blood levels are increased in obesity[2] and following fat consumption;[3] that administration of leptin to *ob/ob* animals—which were never previously exposed to the hormone—causes the mice (as well as genetically normal rodents given large doses) to lose weight[4,5] and become fertile;[6] that proteins that bind leptin ("leptin receptors") are present within various peripheral tissues and, in considerably smaller amounts, within hypothalamus and choroid plexus;[6] and that many scientist believe that leptin constitutes "...an adipose tissue-derived signaling factor for body weight homeostasis..." [*sic*] which reduces food intake by "...direct binding to receptors in the central nervous system."[7] From the moment of leptin's discover, the unassailable fact that adipocytes produce and release this protein has been spun as providing support for a cherished but largely un-

proved theory—that body weight and/or fat content are regulated by some sort of "set-point" mechanism. One unintended consequence of this coupling is that we now know a lot about what leptin does not do (for example, it does not cause weight loss by decreasing food intake, and leptin deficiency does not cause human obesity), but very little about what it does.

The leptin-deficiency model of human obesity was quickly made untenable by the finding that plasma leptin levels are increased in humans (and rodents) made obese by overeating, and that the obesity and overeating persist in spite of these increases.[2,3] So the presumed locus at which the "leptin system" malfunction causes human obesity was shifted: Now the offending protein no longer was leptin itself but another set of gene products, in the brain, which combined with leptin. These leptin receptors were suggested as residing either in hypo-

Reprinted with permission from *Nature Medicine,* Vol. 2, No. 5, May 1996.

thalamic neurons, where they transduced the leptin signal ultimately affecting food intake, or in the choroid plexus, where they served to transport the large leptin molecule across an otherwise insurmountable blood-brain barrier. However, no signal transduction system was shown to be activated when leptin coupled with its "neuronal receptors," leaving their receptor status in doubt, and no explanation was offered as to why these proteins were so much more abundant in peripheral organs than in brain. (Indeed, based on the information now available, one could just as well conclude that "hypothalamic" leptin receptors are localized within the endothelial cells in brain capillaries as in neurons.) Similarly, no evidence was presented that leptin in cerebrospinal fluid rises after fat is consumed, as would be expected if the choroid plexus protein served a transport function. Still, so prevailing is the view that any effect of leptin must be mediated by the brain that the investigators who recently demonstrated its ability to restore fertility in *ob/ob* mice[6] apparently didn't bother to look for leptin receptors in the ovary, nor even to determine whether its administration raised plasma levels of pituitary gonadotropins in their animals (as expected if leptin's fertility-promoting action involved the brain).

The most recent blow to the theory that circulating leptin mediates weight homeostasis, and that it does so by activating a brain receptor coupled to eating behavior, was the finding of Levin et al. that, in genetically normal animals, leptin does not reduce weight by decreasing food intake.[8] Using a classic and simple pair-feeding paradigm, control animals were allowed to consume only as much food per day as other lean mice, which had lost weight after leptin administration, consumed electively. The pair-fed mice failed to lose weight. Hence the weight loss caused by leptin had to have resulted from some metabolic effect of the doses used, and not from an action, via the brain, on eating behavior. Because the leptin doses used were never shown to be physiologic, it cannot be assumed that the weight loss they produced bears any relationship to leptin's actual physiologic role, any more than hydrocortisone's ability to ameliorate the itch of poison ivy tells us anything about its actions when used as a substitution therapy for Addison's disease, or more than the strong hungers produced by excessive, hypoglycemic doses of insulin tell us anything about the satiety produced by endogenous insulin released after carbohydrate consumption. It is still possible that very high doses of leptin will be found to reduce body weight in obese people by inducing a hypermetabolic state, much like the thyroid extracts and dinitrophenol used earlier in this century. However, these earlier agents stopped being used because physicians came to recognize that the induction of hypermetabolism was dangerous.

If we are to learn what leptin actually does, *in vivo*, investigators will have to start performing physiologic experiments in which we examine the consequences of varying plasma leptin levels within their normal dynamic range, or identify the factors that act directly on adipocytes to cause leptin's secretion after fat is consumed. Is most leptin secreted by peritoneal fat cells, perhaps perfused by the nutrient-rich blood of the portal circulation? Or by peripheral adipocytes, perhaps responding to circulating fatty acids, or even to other hormones released after fat consumption? Or perhaps to sympathetic nervous discharge, or to all of these? We also will have to recognize the limitations of extrapolating findings from experimental animals, like *ob/ob* mice, deprived of leptin or some other protein throughout their development, to normal animals or humans. It seems just as likely as not that leptin has a role in development, hence animals whose cells and regulatory systems have never been exposed to the hormone cannot be expected to respond the way "real" mice do when given it in adult life. Similar caveats apply to the use of other mutated or "knockout" mammals to learn about normal physiology.

Now let the search begin for what leptin really does....

References

1. Zhang, Y. et al. Positional cloning of the mouse obese gene and its human homologue. *Nature* **372**, 425-431 (1994).
2. Considine, R.V. et al. Serum immunoreactive-

leptin concentrations in normal-weight and obese humans. *New Engl. J. Med.* **334**, 292-295 (1996).

3. Frederich, R.C. et al. Leptin levels reflect body lipid content in mice: Evidence for diet-induced resistance to leptin action. *Nature Med.* **1**, 1311-1314 (1995).

4. Halaas, J.L. et al. Weight-reducing effects of the plasma protein encoded by the *obese* gene. *Science* **269**, 543-546 (1995).

5. Campfield, L.A., Smith, F.J, Guisez, Y., Devos, R. & Burn, P. Recombinant mouse OB protein: Evidence for a peripheral signal linking adiposity and central neural networks. *Science* **269**, 546-549 (1995).

6. Chehab, F.E., Lim, M.E. & Lu, R. Correction of the sterility defect in homozygous obese female mice by treatment with the human recombinant leptin. *Nature Genet.* **12**, 318-320 (1996).

7. Tartaglia, L.A. et al. Identification and expression cloning of a leptin receptor, OB-R. *Cell* **83**, 1263-1271 (1995).

8. Levin, N., Nelson, C., Gurney, A., Vandlen, R. & de Sauvage, F. Decreases food intake does not completely account for adiposity reduction after ob protein infusion. *Proc. Natl. Acad. Sci. USA* **93**, 1726-1730 (1996). ❑

Questions

1. What is leptin?

2. What effect does leptin secretion have in the body?

3. What kind of model did researchers use to study the role of leptin in the body?

Answers are at the back of the book.

20 *Neuroscientists are claiming the dawn of a new era in dendrite biology. Advanced techniques are revealing that dendrites have a function beyond the computing of synaptic potential that travel across its path. Recent results indicate that signals travel a two-way highway conveying messages to the cell body and then relaying them back to the outer reaches once again. And the signals are not limited to synaptic-potentials alone for now there is evidence that action potentials are being initiated within the dendrites themselves. In addition, the action potential can travel back to the dendrite from the axon, thereby producing a better response to future signals. Called plasticity, this back-propagation of action potentials does strengthen the synapse and demonstrate that dendrites play an important role in actively shaping the responses of neurons and in the formation of learning and memory.*

Dendrites Shed Their Dull Image

Marcia Barinaga

Science, April 14, 1995

New techniques are revealing that dendrites, once thought to be mere adding machines, seem to be more actively involved in shapping the responses of neurons.

What goes on behind drawn curtains in the tight community of the brain? Consider the lacy, branching appendages of nerve fells known as dendrites. Dendrites have long had a reputation as solid but dull citizens of the neural metropolis: bean counters who passively add up the information they receive through the synapses dotting their surface. But it isn't easy to study dendrites directly, and some researchers have suspected that, like what goes on in the banker's basement at midnight, the secret lives of dendrites are more complex and interesting than their public image suggests.

Recent results indicate that those suspicions may be right. In the past few years, neurophysiologists, aided by new techniques that allow direct examina-tion of dendrites, have begun to penetrate the mystery, producing some of the clearest insights yet into how dendrites work. Their studies mark the beginning of a "new era in dendritic biology," according to Columbia University neuroscientist Eric Kandel.

In that new era, researchers are confirming long-held suspicions that dendrites, far from being bean counters, play an active role in shaping the life of the neuron. Their finely branched network acts as a two-way highway, not only conveying incoming messages to the cell body, but also relaying information from the cell body back to their own outer reaches—information that may modify their responses to future signals. This ability of the dendrites to react and adjust their activity is likely to play an important role in such mental processes as learning and memory. The work has "really put the spotlight on the dendrites," says neuroscientist and neural modeler Terrance Sejnowski of the Salk Institute.

The "bean-counter" image of dendrites stems from the fact that their job as information receivers requires that they pass along to the cell body the synaptic potentials, the little dollops of charge that enter the dendrites when neurotransmitter molecules activate synapses. During the journey to the cell body, each synaptic potential adds up with all the others moving through the dendrites. If their sum is large enough when they reach the cell body, they trigger an action potential by causing sodium ions to flow into the cell. The action potential then sweeps down the axon, the part of the nerve cell responsible for carrying the signal to other neurons.

In the traditional dogma of neuroscience, action potentials were limited to axons, and synaptic potentials to dendrites. "This made a clear distinction between the input and the output of the cell," says Salk Institute neurophysiologist Chuck Stevens. But more than 30 years ago, cracks began to appear in that simple model, as researchers reported evidence of action potentials in the dendrites. Some theorists were intrigued by this possibility and began to consider what action potentials might be doing in the dendrites. Others pooh-poohed the idea, saying that action potentials sweeping through the dendrites would foul up their ability to add up synaptic potentials.

The debate was complicated by the fact that dendrites were considered too fine to be impaled easily with electrodes. Their diameter is a mere 1/1000th the diameter of the squid giant axon, which was used for studies of action potentials in axons, and 1/10th the diameter of a typical neuron cell body, which is where neurophysiologists generally put their electrodes. That meant that, at first, all the evidence for reaction potentials in the dendrites was gathered indirectly. In the 1970's however, Rodolfo Llinás's team at New York University poked electrodes directly into dendrites of neurons from the cerebellum and recorded action potentials there. But critics worried that the delicate dendrites may have been damaged by the impaling electrode, producing an erroneous result.

Even those researchers who believed the action potentials were real couldn't agree on which way they were going. Some experiments suggested ac-

tion potentials may initiate within the dendrites themselves. Others suggested they started in the axon and spread into the dendrites from there. The field was clearly in need of new approaches.

A key advance came last year from Greg Stuart, a postdoc with Bert Sakmann at the Max Planck Institute for Medical Research in Heidelberg, Germany. Sakmann shared the 1991 Nobel Prize for developing a technique called patch-clamping, which replaced sharp, impaling electrodes with smooth-ended electrodes which are pressed up against the neuron's outer membrane to form a tight electrical seal. Stuart had developed methods for patch-clamping dendrites that took advantage of new contrast-enhancing microscopy techniques to bring the dendrites sharply into view.

Using the new methods, Stuart put separate patch clamps on the cell body and dendrites of individual neurons in slices of cerebral cortex from rat brains. His results, reported in the January 6, 1994 issue of *Nature*, showed that, when the dendrites were stimulated and the neuron fired, the resulting action potential was registered first by the electrode on the cell body and then, milliseconds later, by the electrode on the dendrites.

This was the first direct evidence that actin potentials travel into the dendrites after being triggered in the axon. "This dual-patch technique that the Sakmann lab introduced really nailed this subject down in a way that wasn't capable of being done in the past," says neuroscientist William Ross of New York Medical College in Valhalla.

The idea of "back-propagating" action potentials (so called because they travel against the main flow of information from dendrite to cell body to axon) had been kicked around among theorists for years. It has a certain appeal, because it could serve as a feedback signal to the dendrites and the synapses on their surfaces.

Researchers have long known that neurons can adjust the strength of their response to incoming signals, a process called plasticity that is thought to be important for learning. For example, in one form of plasticity, synapses that are active when a neuron fires become stronger, so that they give a bigger response to future signals. Back-traveling action potentials seem "like the ideal way to tell the synapses something happened," says neuroscientist Dan Johnston of Baylor College of Medicine in Houston. "How else will they know the cell fired?"

To find out whether the action potentials might in fact have such a role, Sakmann postdoc Nelson Spruston turned to the neurons in which synaptic plasticity has been most intensely studies, the CA1 neurons of the hippocampus, a brain structure involved in some forms of learning. Sakmann's group reports that back-propagation occurs in the CA1 neurons, although they don't yet know if it influence plasticity.

Although they have not answered that question, they made an unexpected discovery that might relate to plasticity. They noticed that sometimes, after a neuron had fired several times in succession, the electrode on the dendrite would register a much-reduced signal, as if the later action potentials in the series fizzled out before reaching it.

Clarification of what was happening came from Sakmann postdoc Yitzhak Schiller, who was studying action potentials by injecting neurons with calcium-sensitive dyes. Action potentials let calcium into the cell, so Schiller could use the dyes to record the path of the action potentials out to the finest tips of the dendrites. He found that when the neuron fired off a series of action potentials in a row, the first one would spread unhindered throughout the dendrites, but subsequent ones would not make it past some of the dendrites' branch points.

When Schiller and Spruston saw each other's result, "we got really excited," Spruston says. "It all fit together very nicely...You have the idea of all these branch points acting like switches that can control the number of action potentials that get through those points." Their finding was confirmed by Joseph Callaway, a postdoc in Ross's lab at Valhalla, who has similar results with calcium-sensitive dyes.

Still, a big question remains unanswered: What is the significance of this switching? If the back-propagating action potentials do contribute to plasticity, says Columbia's Kandel, the fact that branch points can act as switch points means that "one [branch] will [receive] the action potential and one will not.…

You will have the capability in one branch for plasticity that the other branch will not have." That adds to the computing potential of the dendrites, as selective strengthening of some branches would give a boost to signals arising there, leaving other branches unaffected.

Despite their apparent usefulness, backward-traveling action potentials are not universal in dendrites. NYU's Llinás found, in neurons called Purkinje cells, the action potentials begin with the dendrites themselves and travel toward the cell body. But these are not typical action potentials. Purkinje cells lack sodium channels in their dendrites, and the action potentials Llinás observed are carried by calcium ions instead. Indeed, the lack of sodium channels in Purkinje cell dendrites may explain why action potentials that start in the cell bodies of those neurons do not travel backward into the dendrites, a fact that Stuart and Michael Häusser confirmed with dual patch-clamp recording, as reported in a paper in last September's issue of *Neuron*. While the role of the calcium-based action potentials in Purkinje cells is not known, their existence attests to a different "philosophy of [function] of the two cell types," says Llinäs.

Besides resolving the question of whether dendrites carry action potentials, recent research is clearing up another long-standing mystery about dendrite action, and that is how synaptic potentials travel through the dendrites. Action potentials travel "actively" which means that, as they pass through the neuron, they open ion channels, allowing positively charged ions to flow into the cell and add their charge to the traveling signal, like springs continually renewing a river's flow. Synaptic potentials, on the other hand, we thought to travel passively, like creeks in dry country, simply flowing along the inside of the membrane with no additional inputs.

The problem is that a potential traveling in this manner loses charge as it goes, and so synaptic potentials from the most distant dendrites would virtually disappear before reaching the cell body. But researchers long ago showed that those distant synapses are able to fire a neuron, which means those synaptic potentials must be able to make the long journey. "You wonder how could these events on the distal dendrites ever make it," says Yale University neurobiologist Tom Brown. "The obvious answer is there has to be some booster out there."

Studies with ion-sensitive dyes had suggested that there are ion channels in the membranes of dendrites that might act as boosters. Baylor's Johnston and postdoc Jeffrey Magee provide more direct evidence that that is the case. They patch-clamped the dendrites and studied individual channels in the patches of membrane beneath the electrode to find out under what conditions the channels opened. "We allowed the cell to experience normal synaptic potentials," says Johnston, "and asked the question of whether or not, in the little patch of membrane, the channels will open during the normal physiological event." They found that voltage-dependent sodium and calcium channels in the membrane open not only in response to action potentials, but also when synaptic potentials pass by on their way to the cell body. "It's like power lines," says Johnston. "You have transformers and amplifiers along the way" to maintain and modify the signal.

Johnston points out that the entry of calcium ions may also be an important cause of plastic changes in the dendrites. Indeed, calcium is a powerful intracellular signal. Whether let in by back-propagating action potentials or by synaptic potentials, it is bound to pay a role in shaping the future response of the dendrites. "There are likely to be many, many processes triggered by calcium." says neuroscientist Roberto Malinow of Cold Springs Harbor Laboratory, "synaptic potentiation…, regulation of ion channels, and even [dendrite] growth."

Despite all the recent revelations, neuroscientists have a long way to go before they understand how all this activity in the dendrites contributes to mental processes. But the next round of experiments addressing that question is already well under way in many labs. For example, Sakmann's group is testing to see whether back-propagating action potentials are necessary to strengthen synapses, and Sejnowski's team at Salk has a paper in press at the *Journal of Physiology* in which they report that the firing of action potentials in CA1 cells strengthens the cells' response to later signals, possibly through bigger boosts to the synaptic potentials as they travel to the cell body.

Ultimately the same techniques will be applied to many other neuron types and will reveal even more variety in the activities of dendrites. "It is all coming together very rapidly," says Sejnowski. And as it does, the staid image of the dendrite is being replaced by something far more intriguing. ❑

Questions

1. What happens to synaptic potentials that originate in the dendrites?

2. What is the patch-clamp technique?

3. What value do backward-traveling action potentials have in the dendrite?

Answers are at the back of the book.

21 *Chemical synapses connect sensory cells to neurons, neurons to neurons, and neurons to their target organs. In the brain, neuronal activity depends on the neurotransmitter dopamine to signal cells involved in movement, cognition and emotion. Under normal conditions, dopamine levels are tightly controlled but addictive drugs such as cocaine and amphetamines reek havoc on the dopamine-signaling pathwayay by dumping too much of the chemical into the synapse. In addition, cocaine can hamper the removal of dopamine from the area by blocking a special transporter. Researchers are hoping this information will lead to the development of new therapies for drug addiction.*

New Clues to Brain Dopamine Control, Cocaine Addiction

Michael Balter

Science, February 16, 1996

Maintaining normal brain function, like many things in life, requires a delicate balancing act: Too much neuronal activity can be just as bad as too little. Researchers have long known that the chemical dopamine—one of the neurotransmitters that relay signals between nerve cells—is essential in such functions as movement, cognition, and emotion. But how the brain keeps its dopamine levels in equilibrium has been the subject of intense inquiry. Now, a research team led by Marc Caron of the Howard Hughes Medical Institute lab at Duke University has shed new light not only on this balancing act, but also on the mechanisms by which drugs, including cocaine and amphetamine, cause addiction.

The team reports in the 15 February [1995] issue of *Nature* the creation of a strain of "knock-out" mice in which the gene for a protein known as the dopamine transporter has been eliminated. Previous evidence suggested that the transporter, which mops up the dopamine that nerve cells release when they fire, helps regulate dopamine signaling. Many experts also believe that cocaine, and perhaps amphetamines, cause their addictive highs by raising brain dopamine, possibly by acting through the transporter. But Caron and his Duke colleagues Bruno Giros, Mohamed Jaber, and Sara Jones, and collaborator Mark Wightman of the University of North Carolina have shown that the transporter is ever more important than once thought.

They found that the knockout mice, which are unable to clear dopamine from the synapses between nerve cells, become markedly hyperactive, despite heroic attempts by the animals' nervous systems to turn down their dopamine-signaling pathways. Caron and his colleagues conclude that the transporter is the key factor controlling dopamine levels. And this conclusion is buttressed by their finding that the animals are completely unaffected by administration of cocaine and amphetamines, which suggests that these drugs exert most of their effects through the transporter. "We have demonstrated for the first time the really primordial role it plays," Caron says.

Other experts agree. "This is a wonderful paper," says Michael Kuhar, a neuropharmacologist at the Yerkes Regional Primate Research Center at Emory University in Atlanta. "It's going to be a classic in the field." And Alan Leshner, director of the U.S. National Institute on Drug Abuse (NIDA), says the new mouse is "a fantastic tool" to help understand addiction, because "every drug of abuse works through the dopamine system" to a significant extent. Indeed, the work may help in the design of new drugs for treating drug addiction, and perhaps also for Parkinson's disease, which is caused by a sharp fall in brain dopamine concentrations.

Evidence that the transporter is important in regulating dopamine has been building for some time. Researchers had found, for example, that it captures the dopamine released into synapses and then pumps it back into the nerve cell, where it is repackaged in storage vesicles. But because the transporter is only one of several elements of the dopamine system, Caron's group used the knockout approach to see just how key the transporter's role is.

The results were dramatic: Mice lacking the transporter were five to six times more active than "wild-type" mice who have the gene, as measured by the number of times they passed through photocell beams in an "activity box." By itself, this hyperactivity was not surprising: Without the transporter, dopamine was expected to accumulate in the synapse and continue to activate nerve firing. However, the researchers were surprised to find much less dopamine was produced in the knockout's brains compared to normal mouse brains.

To explain this seeming inconsistency, the Duke team demonstrated that the animals try to compensate for the lack of the transporter by "down-regulating" the entire dopamine system. Not only did their brain neurons release less dopamine when stimulated, but they also made less of a key enzyme involved in dopamine syntheses. And at the other side of the synapse, there was a dramatic decrease in the number of the receptors through which dopamine exerts its effects. "The animals are doing everything possible to dampen the signal generated by absence of dopamine reuptake," Caron says. But these efforts weren't enough: In the knockout mice, dopamine remains in the synapse 100 times longer than in wild-type mice, producing an enhanced effect despite the much lower concentrations.

The group's finding that the knockout mice lost their sensitivity to cocaine and amphetamines is consistent with recent work indicating that cocaine raises dopamine concentrations in synapses by binding to the transporter near the site where the neurotransmitter itself binds, thus blocking dopamine reuptake. But, says NIDA's Leshner, "we've never had a technique like this, where you can see just how necessary the transporter is to the psychostimulatory effect." As for amphetamines, earlier studies had suggested that these drugs acted primarily by increasing dopamine release—a view that might now have to be modified. The Duke workers "show that the transporter is a mandatory component for dopamine release triggered by amphetamines, and that's a surprise," says neurobiologist Ann Graybiel of the Massachusetts Institute of Technology.

Caron's team is launching a new series of experiments to fine-tune their understanding of the dopamine transporter's role. For example, Giros, now at the French biomedical agency INSERM in Paris, will lead a study to see if the knockout mice self-administer cocaine and amphetamines. "That will be a key experiment," says Kuhar. If the Duke workers are right about the dopamine transporter, mice lacking this protein would be unlikely to derive any pleasure from these drugs and thus won't self-administer them.

If this prediction pans out, it might boost efforts to develop new therapies against drug addiction, particularly cocaine dependence, which Leshner says is a major NIDA priority. This might be done, he suggests, by designing drugs that prevent cocaine from binding to the transporter. And a greater appreciation of the role of the transporter could also lead to new medications for Parkinson's disease; the symptoms of low dopamine might be alleviated by blocking the transporter. "This paper shows you can do a lot with a little bit of dopamine, if you can just keep it in there," says Graybiel. ❑

Questions

1. What role do neurotransmitters play in brain function?

2. How does dopamine exert its effect?

3. What did researchers find as a key factor for controlling dopamine levels in mice given coucaine and amphetamines?

Answers are at the back of the book.

Section Four

The Mechanical Forces and Chemistry of the Senses

Scientists are continuing to find novel ways in which nitric oxide is used in the body. The newest finding is right under your nose, or more precisely, in the sinuses where it protects them from infection. As a gas, nitric oxide binds to vital enzymes in bacteria, thus stopping their growth and reproduction. In addition, doctors are capitalizing on this molecule's ability to vasodilate blood vessels and increase oxygenated blood flow in the lung by pumping air from the own nose directly into the patient's hospital ventilator. This could improve the blood pressure in certain patients without the worry of administering too much nitric oxide externally.

NO in the Nose
Discover, July 1996

Unlike the nose, which harbors many types of bacteria, the sinuses—the air-filled cavities in the skull surrounding the nose—remain curiously free of intruders. The sinuses, like the rest of the body, are a warm, bacteria-friendly environment. "It should be perfect for bacteria to live there," says Jon Lundberg, a physician at the Karolinska Institute in Stockholm. Lundberg and his colleagues have discovered one way that the sinuses stay sterile: they produce nitric oxide, a gas considered a pollutant in the atmosphere and one that is lethal even in small doses to bacteria and viruses.

Lundberg and his colleagues used a syringe and a catheter to remove air from the sinuses of dozens of healthy, nonsmoking subjects. When they analyzed the air samples, they found nitric oxide levels that were close to the highest permissible atmospheric pollution levels. They later removed small pieces of the membrane lining the sinuses, which consist largely of epithelial cells, and found that these cells were the source of the nitric oxide. Nitric oxide, says, Lundberg, binds to vital enzymes in microbes, stopping their growth and reproduction and thus protecting the vulnerable sinuses from infection.

But NO has another important role, Lundberg says, one that doctors have known about for some time: it is a powerful vasodilator, a substance that dilates blood vessels. Since the sinuses create high concentrations of NO, when we breathe through our noses, NO probably travels down the airway to the lungs. By then it is so diluted that it isn't harmful—and according to Lundberg may even be beneficial: it may dilate blood vessels and increase blood flow specifically in areas of the lung that are well supplied with oxygen. That would increase the amount of oxygen in the blood. "This is kind of a new physiological principle," says Lundberg.

Doctors already use NO to open up the blood vessels of patients on ventilators suffering from high blood pressure in the lungs. Lundberg thinks it could also help such patients in another way. Typically, ventilator patients receive air directly from their trachea, so they miss out on the NO being generated in their sinuses. In a small trial, Lundberg has found that he can increase the oxygen content of his patients' blood by nearly 25 percent by simply pumping their own nasal air into their ventilators. Not only is the treatment effective, but it is also less risky than externally administered NO might be. "You can never give too much," explains Lundberg, "because you only give back what is taken away." ❏

Questions

1. How does nitric oxide work in the sinuses?

2. What is nitric oxide's most important role in the body?

3. How can nitric oxide from the sinuses be used to help other parts of the body?

Answers are at the back of the book.

---

23 *There's good news for sufferers of glaucoma, a disease which can cause vision loss. Ophthalmologists are using the advanced technology of lasers to help relieve the intraocular pressure in the eye. By treating the trabecular meshwork of the eye with argon lasers, the cells that reside there start secreting enzymes which change the matrix making it easier for intraocular fluid to exit. The benefits of laser surgery are promising over drug therapy and conventional mechanical surgery, which both carry significant risk of side effects. While some risk is still associated with this new treatment, laser surgery could still preserve the eyesight of many patients who have found little help elsewhere.*

Lasers Win Out in Glaucoma Trial

Norman Bauman

New Scientist, January 6, 1996

Lasers are as good as drugs at treating glaucoma, according to a long-term study in the US. This should dispel doubts about the effectiveness of lasers, even though the mechanism of glaucoma, and the reason why lasers are so effective, are still poorly understood.

Glaucoma is caused by a build-up of pressure in the eye, which destroys nerve cells and causes loss of vision. Conventional surgery to treat the condition cuts a channel in the eye to drain excess fluid but the procedure carries a 5 percent risk of significant loss of vision. Several drugs can reduce the pressure, but they can cause serious side effects such as raised blood pressure and bronchospasm. Laser surgery seems to carry no risk of major side effects, and this study shows that it is as effective as preserving eyesight as drug treatment.

The study was carried out at several centres around the US, and the results were reported in the *American Journal of Ophthalmology* (vol 120, p 718). The normal pressure in the eye is around 2000 pascals.

The 203 patients who took part in the study started out with a mean intraocular pressure of around 3500 pascals, according to Jacob Wilensky, an ophthalmologist at the University of Illinois, Chicago. Each person was treated in one eye with eye drops, and in the other with an argon laser. Three months later, the mean intraocular pressure had dropped to 2400 pascals.

For most of the participants, the doctors were able to maintain the lower pressure and stable visual field for an average of 7 years. But 10 percent of patients need conventional surgery.

"We could demonstrate that the laser has a statistically significant benefit over medication," said study chairman Hugh Beckman, of Sinai Hospital, Detroit. "But we couldn't demonstrate that that small benefit was clinically significant." The eyes treated with the laser had 160 pascals lower intraocular pressure, and slightly lower loss of vision.

Glaucoma is often a disease of old age, and older patients may find it particularly difficult to tolerate the side effects of drugs. Wilensky says that a 75-year-old patient with newly diagnosed glaucoma could have the condition controlled for the rest of his life by a single laser treatment.

Wilensky says that neither laser treatment nor medication can be guaranteed to keep working after a few decades, though some ophthalmologists disagree. On the other hand, mechanical surgery, while riskier, can permanently lower the pressure to whatever level is required. Wilensky says that a 45-year-old patient will probably need conventional surgery eventually, so doctors may as well operate immediately.

Intraocular fluid is secreted behind the iris, and flows past the pupil into the front chamber of the eye. It normally flows out through the trabecular meshwork, a ring of permeable tissues at the periph-ery of the iris where it joins the cornea. Glaucoma can occur when the meshwork gets clogged.

Surgeons started using lasers in 1979 to cut a hole in the trabecular meshwork to drain the fluid. The procedure worked, but not for the expected reason. Instead of simply drilling a hole, the laser stimulated a reaction in the mesh that changed its chemistry and made it more permeable.

The matrix that holds the cells in the trabecular meshwork together is constantly destroyed and recreated, and an imbalance could clog it up, says Ted Acott, of the Oregon Health Sciences University, Portland. He says that laser treatment causes trabecular cells to secrete enzymes that disrupt the matrix, so the laser appears to affect the biochemistry of the eye, rather than its plumbing. ❑

Questions

1. What is glaucoma?

2. What is the normal pressure in the eye?

3. How does laser treatment help relieve the intraocular pressure in the eye?

Answers are at the back of the book.

24 *Exciting new discoveries are being made in the area of hearing research and scientists are looking to our fine feathered friends for more answers. Studies with neonatal chicks have found that hair cells, the receptor cells in the cochlea of the inner ear, are capable of regeneration. Not only are new hair cells restored to the cochlea but they are formed with functional connections to the brain. In addition, vestibular organs used in balance also show damaged-induced regeneration. With the development of an in vitro model, it will become easier to identify the specific molecules needed to promote a regenerative response in the mammalian hair cell, which may prove clinically important in curing hearing loss and balance disorder in humans.*

Stimulating Hair Cell Regeneraton: On a Wing and a Prayer

Edwin W. Rubel and Jennifer S. Stone

Nature Medicine, October 1996

Identifying the mechanism that regulates hair cell regeneration may contribute to developing treatments for hearing loss in humans.

It can be said that the most important attribute humans possess is the ability to communicate—language and its associated sensory-motor systems are at the heart of all human progress. Loss of hearing due to genetic or environmental causes is devastating, affecting the lives of hundreds of millions of people worldwide and is ubiquitous among the aged. The findings reported by Navaratnam *et al.*[1] provide insight into the regulation of auditory receptor epithelial regeneration and are an important contribution to the struggle to find a cure for hearing loss. The authors show that the regeneration of inner ear hair cells in birds involves activation of the cAMP/PKA (protein kinase A) intracellular signaling pathway. Using *in vitro* preparations of the chick inner ear they examined the role of this

signaling pathway in two ways. First, in intact tissue, they showed that addition of cAMP accumulators markedly increased the rate of supporting cell proliferation and that this effect was blocked by inhibitors of PKA, the major effector of the cAMP pathway. Second, hair cell regeneration in response to aminoglycoside damage was shown to be reduced in the presence of PKA inhibitors. Finally, the authors demonstrated that cAMP-stimulated mitotic activity resulted in the production of new cells with the phenotypic characteristics of hair cells.

Hearing loss is usually due to the death of hair cells, the receptor cells in the cochlea of the inner ear. In humans and other mammals, hearing loss is permanent because these cells are not spontaneously replaced. Treatments are limited to sound amplification using hearing aids, which unsatisfactorily attempt to overstimulate the surviving hair cells or, in the cases of profound hearing loss, the direct electrical stimulation of auditory nerve endings with an implanted electrode (cochlear implant) that bypasses the receptor cells. Although these treatments help many people, they do not solve the fundamental problem—loss of hair cells—and they cannot restore the subtle elegance of acoustic signals critical for human communication. Thus, textbooks read: "There is no cure for sensory-neural hearing loss."

Hair cells are not unique to the cochlea of mammals; they occur in the lateral line organs and the inner ear balance organs of all vertebrates. It has been known for more than half a century that aquatic vertebrates possess the capacity to produce them throughout life or regenerate new ones under special circumstances.[2] But, until 1987, few people suspected hair cells could regenerate in animals that hear airborne sounds with the sensitivity of mammals or birds. Independent, coincidental, and almost accidental discoveries by two groups conclusively demonstrated that hair cells in the cochleas of neonatal chicks, destroyed by loud noise or drugs, can regenerate.[3,4] This finding opened the door to a new era of hearing research and illustrated the potential of the avian system for furthering our understanding of hearing in mammals.

A flurry of research next showed that the regenerative response in birds involves the production and differentiation of new cells and is not limited to neonatal animals.[5,6] Subsequent studies reported that new avian hair cells restore the cochlear array and form functional connections with the brain, that ongoing hair cell production and damage-induced regeneration occur in the balance (vestibular) organs, and that new hair cells emerge from renewed mitotic activity or direct transdifferentiation (conversion) of underlying support cells.[7] The rapid development of in vitro models of avian and mammalian inner ear epithelia provides opportunities for answering the major questions raised by these earlier studies—which molecules are responsible for stimulating hair cell regeneration in birds and is regeneration possible in the mature mammalian inner ear?[8,9]

The research strategies for stimulating mammalian hair cell regeneration seem obvious: identify the molecules that promote or inhibit proliferation and differentiation in birds and use them to induce a more robust regenerative response in mammals, or discover new factors that promote mammalian hair cell regeneration. In the avian cochlea, damage or death of hair cells appears to play a role in triggering hair cell regeneration, while in the avian vestibular organs the low level of continual turnover of hair cells increases after hair cell damage. The situation in mature mammals is quite different because there appears to be no ongoing cell division in the mammalian auditory or vestibular sensory epithelia in vitro. However, damage promotes a low level of mitotic activity in the vestibular epithelia, demonstrating that progenitor cells are present and that production of new (or recovery of existing) sensory cells is possible.[10-13] In the organ of Corti, there is emerging evidence that new hair cell production may be initiated in vivo after damage.[12] Several laboratories have used in vitro preparations to demonstrate that certain growth factors (for example, transforming growth factor-α) promote cell proliferation in mammalian vestibular organs.[14,15] This approach has been highly productive, and ongoing studies aim to identify novel cytokines and morphogens that stimulate proliferation of the avian and mammalian inner epithelia.

The approach taken by Navaratnam and colleagues is rather different but has yielded important results that begin to define the intracellular pathways regulating the proliferative response of support cells in the avian cochlea. It should be relatively straightforward to test the role played by the cAMP/PKA signaling pathway in the triggering of cell proliferation in the mammalian inner ear. Moreover, these authors have provided a potentially interesting strategy which may prove clinically important: that of bypassing the intercellular communication step and directly stimulating a regenerative response via the cAMP/PKA (or other) signaling pathway.

Obviously, it is a giant leap from understanding regeneration in the avian ear to curing hearing and balance disorders in humans. However, increased understanding of avian hair cell regeneration provides one road toward achieving a goal, that just a decade ago was deemed impossible. By following this and all other available roads, we may find a resolution for those who suffer the social isolation of lost communication.

References

1. Navaratnam, D.S., Su, H.S., Scott, S. & Oberholtzer, J.C. Proliferation in the auditory receptor epithelium mediated by a cAMP-dependend signaling pathway. *Nature* Med. **2**, 1136–1139 (1996).
2. Stone, L.S. Further experimental studies of the development of lateral line sense organs in amphibians observed in living preparations. *J. Comp. Neurol.* **68**, 8383–115 (1937).
3. Cruz, R.M., Jambert, P.R. & Rubel, E.W. Light microscopic evidence of hair cell regeneration after gentarnicin toxicity in chick cochlea. *Arch. Otolaryrgol. Head Neck Surg.* **13**, 1058–1062 (1987).
4. Cotanche, D.A. Regeneration of hair cell stereociliary bundles in the chick cochlea following severe acoustic trauma. *Hear. Res.* **30**, 181–195 (1987).
5. Corwin, J.T. & Cotanche, D.A. Regeneration of sensory hair cells after acoustic trauma. *Science* **240**, 1722–1774 (1988).
6. Ryals, B.M. & Rubel, E.W. Hair cell regeneration after acoustic trauma in adult *Coturnix* quail. Science **240**, 1774–1776 (1988).
7. Contanche, D.A., Lee, K.H., Stone, J.S. & Picard, D.A. Hair cell regeneration in the bird cochlea following noise damage or ototoxic drug damage: A review. *Anat. Embryol,* **198**, 1–18 (1993).
8. Oesterle, E.C., Tsue, T.T., Reh, T.A. & Rubel, E.W. Hair-cell regeneration in organ cultures of the postnatal chicken inner ear. *Hear. Res* **70**, 85–108 (1993).
9. Warehol, M.E., Lambert, P.R., Goldstein, B.J. Forge, A. & Corwin, J.T. Regenerative proliferation in inner ear sensory epithelia from adult guinea pigs and humans. Science **259**, 1619–1622 (1993).
10. Forge, A., Lin, L., Corwin, J.T. & Nevill, G. Utrastructural evidence for hair cell regeneration in the mammalian inner ear. *Science* **259**, 1616–1619 (1983).
11. Rubel, E.W., Dew, I.A. & Roberson, D.W. Mammalian vestibular hair cell regeneration. *Science* **267**, 701–707 (1995).
12. Lenoir, M. & Vago, P. Morphological indications of hair cell neodifferentiation in the organ of Corti of amikacin treated rat pups. *Life Sci.* **319**, 269–276 (1996).
13. Tenyen, H., Lopez, I. & Honrubia, V. Histological evidence for hair cell regeneration after ototoxic hair cell destruction with local application of gentamicin in the chinchilla crista ampullaris. *Hear. Res.* **89** 194–202 (1995).
14. Lambert, P.R. Inner ear hair cell regeneration in a mammal: Identification of a triggering factor. *Laryngoscope* **104,** 710–718 (1994).
15. Yamashita, H. & Oesterle, E.C. Induction of cell proliferation in mammalian inner-ear sensory epithelia by transforming growth factor alpha and epidemal growth factor. *Proc. Natl. Acad. Sci. USA* **92**, 3152–3155 (1995). ❏

Questions

1. What causes hearing loss?

2. What is the new strategy for understanding the regulation of auditory receptor epithelial regeneration?

3. When damaged, can hair cells spontaneously regenerate?

Answers are at the back of the book.

25 *Imagine not having to wear glasses or contact lens ever again. Corneoplasty is a new procedure developed by ophthalmologists to correct conventional eye defects in patients with short and long-sightedness. A transparent mold is created, much like a ordinary contact lens which when worn for several months reshapes the cornea and corrects the defects. So far, the technique appears to be successful as vision is improved with less surgery than radial or laser keratectomy treatments. Don't throw away the spectacles just yet as results from product testing in US hospitals must prove this is a practical ophthalmic therapy for correcting vision before its released on the market.*

I Spy with My Moulded Eye

Andy Coghlan

New Scientist, March 16, 1996

The ophthalmic equivalent of a dental brace may offer a cure for common sight defects. Custom-built transparent moulds temporarily worn by patients could reshape the cornea—the disc at the front of the eye that refracts light onto the lens. The idea is that once the mould is removed, the cornea retains its new shape for good.. Such a procedure could eliminate the need for spectacles or contact lenses many people, say the developers.

The moulds have been tested in 15 patients with common visual defects in Mexico. "Everyone's vision got better or stayed the same, and no one's vision got worse," says Diana Schmidt, the vice-president of marketing at Advanced Corneal Systems (ACS), the company in Laguna Hills, California, that developed the technique. She says that the earliest patients with conventional eye defects still have improved vision 17 months after undergoing the procedure, which suggests that the corneas will retain their new shape.

The only surgery required is a single injection of a drug into the rim of the cornea. In contrast, radial kerototomy involves making multiple incisions into the cornea with a scalpel, while laser photorefractive keratectomy relies on excimer lasers to remove slivers of the cornea.

ACS's procedure, called corneoplasty, first requires ophthalmologists to work out how a patient's cornea needs to be remoulded. Schmidt says that machines for mapping the contours of the cornea, called corneal topographic analysers, are already available in many eye treatment centres and are often used as a prelude to other forms of ophthalmic treatment.

By collecting reflections from rings of light shining onto the patient's eye, the machine gradually builds up a contour map of the cornea. In corneoplasty, this map is fed into a machine, which custom-builds a transparent mould for correcting any defects. The mould is made from the same material as ordinary contact lenses.

Short-sighted people, who can only see nearby objects clearly wear moulds that flatten their corneas. The flatter corneas help the lens focus light from distant objects on the retina, instead of in front of it.

Long-sighted people, who can only see distant objects clearly, wear moulds that make their corneas more rounded. "You want the cornea to bulge out a little, so we press it in slightly from the edges," says Schmidt. This brings the focus onto the retina from behind the eye.

Before fitting the moulds a drug containing an enzyme that softens the cornea is injected, making it easier to remould. The enzyme temporarily breaks chemical bounds in proteoglycans—sugary molecules that hold fibres of transparent collagen rigid in the cornea. According to Schmidt, these bonds repair completely after three or four months. The company will not reveal details about the enzyme or the drug, but says that the drug has been used safely in eye surgery for years.

For the first three days, patients wear the mould throughout the day. Thereafter, they wear the moulds only when they sleep. Once the treatment is complete, after around three months, patients can dispense with the moulds for good. Schmidt says that the moulds are painless to wear. "They are larger than normal contact lenses, covering the en-tire cornea almost to the white of the eye," she says.

The first safety trial was in blind people, and the second, tentative trial tested the procedure on people with incurable vision defects. Only the final 15 patients had the conventional sight defects that the company hopes to treat, and all reported improved vision as a result.

Schmidt says that further trials at 10 eye hospitals in the US will start soon. They will be coordinated by Anthony Nesburn, an eye surgeon at the Cedars-Sinai Medical Center in Los Angeles.

"Individual ophthalmologists will set the price after buying the drug and the lenses from us," says Schmidt. Existing techniques cost up to $2000 per eye.

ACS has submitted its findings for publication, and is to present the full results next month in Florida at the annual meeting of the American Association for Vision and Research in Ophthalmology.

Cornea specialists remain skeptical however. "Any injection into the cornea is potentially dangerous," says Charles McGhee, ophthalmic surgeon and professor of ocular therapeutics at the Eye Infirmary in Sunderland. He says that it could carry a risk of infection or scarring. ❑

Questions

1. How do ophthalmologists design a custom corneal mold for each individual?

2. How should the cornea be changed to correct the vision for short-sighted people?

3. How should the cornea be changed to correct the vision for long-sighted people?

Answers are at the back of the book.

26

What makes fingertips sensitive enough to read Braille? There are an estimated 17,000 mechanoreceptors within the palm and fingertips that contribute to the tactile sensation. The mechanism behind that sensory capability is dependent on ion channels that are sensitive to mechanical stimulation. Mechanosensitive channels respond to a tactile stimulus by causing a change in the membrane's ion conductance resulting in the depolarization of the membrane. Researchers study their molecular mechanism using pharmacological agents that act selectively on the channel and recording the electrical response using a pressure clamp device similar to the original patch-clamp. Never before have ion channels been so clearly defined in explaining tactile sensory functions as we know them.

Mechanoreceptive Membrane Channels

Owen P. Hamill and Don W. McBride, Jr.

***American Scientist**, January/February 1995*

Mechanically sensitive membrane channels may participate in processes as diverse as volume regulation in cells and sound reception in vertebrates.

All living organisms appear to have in common the ability to sense and react to mechanical stimulation. So-called mechanosensitive responses range from simple behaviors in unicellular organisms—including a *Paramecium* that reverses its swimming motion after bumping into a barrier—to the more complex behavior of multicellular organisms—including a nocturnal scorpion that is alerted selectively to ground vibrations less than one-billionth of a meter in amplitude. In people, the estimated 17,000 mechanoreceptors in the palm and fingertips of a hand indicate the fundamental importance of tactile sensation. In fact, the refinement of the hand's tactile-grasping ability allowed tool making and most likely promoted the rapid burst in human evolution.

Beyond an organism's ability to sense external mechanical forces, individual cells must be able to sense internal forces that may effect their size and geometry. For example, even a primitive single-cell organism needs the ability to regulate its cell volume when the ionic concentration, or osmolarity, of surrounding fluids varies. Without such volume regulation, an aqueous environment diluted with heavy rainfall, for example, would force a cell to take up water, swell and possibly rupture. This fundamental regulatory mechanism probably arose early in organisms and was conserved during subsequent evolutionary modifications.

Every regulatory mechanism requires an appropriate detection system. In the case of osmotic gradients, a detection system could be based on sensitivity to membrane tension during cell swelling, which could then be used to modulate ionic movement across the membrane. This form of mechano-sensitivity, apart from serving a regulatory function, may also have been modified to sense external mechanical stimuli that deform or stretch a cell membrane. Thus they may underlie sensory functions as we know them.

Although the properties of mechanosensory processes and behaviors have intrigued biologists and physicists for well over a century, direct insight into the molecular mechanisms has only been achieved in the last decade. In particular, high resolution electrophysiological measurements have revealed specific ionic channels in membranes that are gated—opened or closed—by mechanical stimulation. These "mechanogated" channels have been identified in bacteria, fungi, plants and animals. This article focuses on the molecular mechanisms that form a link between a mechanical stimulus and the resulting electrical signal.

From Prods to Pressure Clamps

Initial studies of mechanosensitivity focused on specialized receptors. Such studies revealed that mechanoreceptors can be either *phasic* or *tonic* in their response to stimulation. Phasic receptors—such as pacinian corpuscles that sense vibration in vertebrate skin and vertebrate hair cells that, for one thing, relay sound waves from the middle ear to the nervous system—respond best to transient stimuli. They stop responding, or adapt, when exposed to constant stimulation. Adaptation maintains the high dynamic sensitivity of a mechanoreceptor but avoids saturation from constant background signals, such as gravity in the case of hair cells that signal head movement. Tonic receptors—such as Merkel cells in mammalian skin and slow-adapting stretch receptors in crustacean muscle—report constant stimulation, and they show little or no adaptation. These characteristics place constraints on the cellular and molecular mechanisms behind mechano-electrical transduction, in which a mechanical event is transcribed into an electrical, or ionic, event that serves as information for the nervous system.

Classical studies of mechano-electrical transduction required a probe to prod the receptor and an intracellular electrode, a fine wire or glass capillary,

to record the electrical response of the entire cell. In pacinian corpuscles, for example, mechanical stimulation leads to a transient increase in a receptor membrane's conductance to sodium and potassium ions, and somewhat to calcium ions. In a resting receptor, one not responding to a stimulus, sodium ions are concentrated in the intracellular fluid, and potassium ions are concentration in the intracellular fluid. Because a resting nerve-cell membrane has a higher permeability to potassium ions compared with sodium ions, the ionic contraction differences result in a net efflux, or outward flow, of potassium ions from a cell, which generates a positive charge on the outside of a cell and a negative charge on the inside, or a voltage across a membrane. When channels in the membrane open to both sodium and potassium, the result is a net inward current that decreases the voltage across the membrane, or *depolarizes* it.

Opening channels to sodium and potassium ions depolarizes a membrane though a combination of chemical and electrical forces. Potassium ions tend to move out of a cell to their region of lower concentration, but the positive voltage outside a cell repels the positively charged potassium ions. Positively charged sodium ions, on the other hand, are pulled inward because of the lower concentration and by the negative voltage inside a cell. When channels open to both sodium and potassium ions, the combined chemical and electrical forces create an unbalanced ionic flow: The entering sodium ions outnumber the exiting potassium ions. This creates a net inward current, and the positively charged sodium ions offset some of the negative charge inside the cell, thereby depolarizing it.

In principle, a variety of ion-transport processes could mediate the conductance increase to sodium and potassium ions. For instance, a membrane pump could bring in sodium ions and expel potassium ions. Such pumps, however, are relatively slow and produce small currents. Many mechanoreceptors activate rapidly (in less than one-thousandth of a second) and produce nanoamperes (billionths of an ampere) of current. Such fast-acting, large-amplitude currents are more consistent with gated membrane channels, which can be opened or closed on a millisecond time scale and have transport rates as high as 10 million ions per second. Although gated ion channels were first hypothesized in the 1950s to explain bioelectrical transmission, only the development of the patch-clamp technique in the mid-1970s by Erwin Neher and Bert Sakmann of the Max Planck Institute (for which they were awarded the Nobel Prize in 1991) directly demonstrated individual gated channels in biological membranes. In the patch-clamp method, a glass "patch" pipette is pressed gently against a cell membrane, and suction draws a membrane patch, about two-millionths of a meter across, into the pipette's tip. After a membrane patch seals to the glass pipette, it is isolated electrically from the rest of the cell so that minute currents (less than picoamperes, or trillionths of an ampere) can be recorded in a patch.

Depending on the type of cell and the channel of interest, appropriate stimulation can be applied to a patch to activate specific ion channels in a membrane. For example, voltage-gated channels, those that open and close when exposed to a specific voltage, can be activated by controlling the voltage of a patch of membrane; specific chemicals can be included in the patch-pipette solution to activate transmitter-gated channels, those that are gated by chemical; or suction or pressure can be applied to a patch to activate mechanogated channels. In the original patch-clamp technique, suction or pressure was applied by mouth or syringe. More recently, we developed a pressure-clamp arrangement that provides the rapid and precise suction or pressure waveforms that are necessary to describe the dynamic properties of mechanogated channels.

Tapping Toad Eggs
Ideally, one would like to identify and characterize single mechanogated channels in sensory mechanoreceptors such as pacinian corpuscles or hair cells. So far, however, these specialized mechanoreceptors have proved inaccessible to patch-clap studies because they are either embedded deeply in tissue, surrounded by accessory structures, or have specializations, such as cilia, that prevent direct sealing of a pipette on their mechanosensitive membrane. Fortunately, a wide variety of nonsensory cells, which

are accessible to patch-clamp recording, also express mechanogated channels.

The first indication of mechanogated channels in nonexcitable, or non-neuronal, cells came from patch-clamp studies of volume regulation in red blood cells of frogs. Hamill showed that osmotic swelling in these cells activates specific potassium- and chloride-ion-selective membrane channels. Soon after that discovery, Falguni Guharay and Frederick Sachs of the Sate University of New York at Buffalo demonstrated that stretching the membrane of skeletal muscle, via suction on a pipette, activates channels for positively charged ions. Since these initial reports, a variety of stretch-and-volume-sensitive ion channels have been found in the membranes of cells from organisms in all the living kingdoms, and these channels exhibit variability in conductance, ion selectively and gating sensitivity (whether they are turned on or off by mechanical stimulation). These different properties presumably reflect the different physiological functions of the channels.

A particularly convenient preparation for patch-clamp studies of mechanogated channels in nonsensory cells exists in oocytes, female gametes or reproductive cells, of the toad *Xenopus*. This very large single cell, about one millimeter in diameter, possesses an even distribution of mechanogated channels over the cell surface, leading to about 10 million channels. This channel displays a number of properties that are similar to the mechanosensitive conductances in specialized mechanoreceptors, such as pacinian corpuscles. Although the functional role of a mechanogated channel in oocyte physiology has not been determined, it may play a role in sensing membrane tension during volume regulation, cell growth regulation or embryogenesis or perhaps even detecting mechanical perturbations of the membrane by sperm during fertilization.

Opening, Closing and Blocking
An ion channel is characterized in large part by two features: open-channel properties—including conductance, selectively and channel blocking by ions and drugs—and gating properties—including the probability of being open or closed and the conditions that alter that probability. In principle, a chan-

nel could be mechanosensitive because either its conductance or its gaining is altered by mechanical stimulation.

These two possibilities can be distinguished by a pressure-ramp protocol, the application of pressure that increases smoothly in the shape of a ramp. In a patch containing many channels, a pressure ramp generated an electrical response that saturates, or reaches a maximum value, before the pressure reaches its peak level. This indicates that the mechanosensitive current is not induced mechanically as a nonspecific leak or because of damage to the cell. (In fact, the membrane current returns to normal levels after removing the pressure stimulus.) Furthermore, applying the same protocol to smaller patches, containing fewer channels, reveals single events than underlie mechanosensitive currents. These events exhibit two states, either open or closed, and switching between states provides mechanosensitivity. The conductance, or current, associated with a single event remains unchanged as pressure increases. So saturation of a response in a multi-channel patch arises from a dynamic equilibrium between the opening and closing of a finite number of channels. Like many mechanoreceptors, the toad-oocyte mechanosensitive channel allows both sodium and potassium ions to pass when it is open.

One difficulty in studying mechanogated channels has been the relative lack of specific pharmacological agents that act selectively on them. Discovering such drugs could prove crucial to revealing the function of these channels in different types of cells. Amiloride and its various structural analogues, however, block mechanosensitive channels in oocytes that are open. Interestingly, amiloride also blocks fertilization, volume regulation and cell proliferation in various cell types, as well as mechanotransduction in hair cells.

Vibration and Adaptation
Different mechanisms may confer mechanosensitivity on a channel. These mechanisms may be classified as direct or indirect, depending on how the mechanical energy is coupled to the channel-gating mechanism. Direct-coupling mechanisms arise from intramolecular arrangements of the channel molecule and viscoelastic elements inside a cell; biochemical reactions and diffusion do not limit the speed of such processes, which may develop in less than milliseconds. In indirect-coupling mechanisms, by contrast, the channel itself is not mechanosensitive. Instead, a second messenger—such as calcium ions, cyclic adenosine monophosphate (cAMP) or fatty acids—diffuses from a mechanosensitive enzyme to the channel, thereby invoking a longer latency, from several milliseconds to seconds. Using a pressure clamp, we found that the oocyte mechanosensitive channel can turn on in as few as 100 microseconds, which is comparable to the fast turn-on time in vertebrate hair cells. Such rapid activation behavior in an oocyte and a hair cell is more consistent with direct mechanical-gating mechanisms.

In many cases, the turn-off time of a channel is just as important as the turn-on time. These two times, in combination, determine a mechanotransducer's ability to follow oscillations in mechanical energy, such as vibrations. Although oocyte mechanogated channels turn off a little slower than they turn on, the channel should be able to follow mechanical stimulations up to a frequency of about 200 hertz. It is not clear when an oocyte might experience such high-frequency mechanical oscillations, but one possibility is during fertilization. During that process, sperm make a high-frequency bombardment on an oocyte, and an oocyte undergoes rapid changes after a single sperm enters, thereby preventing polyspermy. Mechanosensitive channels may be involved in this process, but little is known presently about the mechanically induced events associated with fertilization.

Beyond high-frequency vibration, a mechanoreceptor may experience a constant stimulus. Like pacinian corpuscles and hair cells, mechanogated channels in both oocytes and skeletal muscle adapt rapidly to a sustained stimulus. Both suction and pressure produce similar levels of adaptation in oocyte mechanosensitive channels, suggesting that a change in membrane tension, not pressure *per se*, activates the channel. Moreover, the channels do not become inactivated or desensitized by a constant stimulus. A mechanosensitive channel adapted to constant suction can be reactivated by increasing the suction.

A simple model suggest a possible mechanism behind adaptation in a mechanosensitive channel. Imagine that a channel is held closed by an elastic element, such as a spring, connected in series with a viscous element, such as a dashpot, which is basically a shock absorber. When a channel is closed, the spring is coiled rather tightly and the dashpot's piston is bottomed out, or pressed against the bottom of the pot. When a channel opens, the spring stretches, and the dashpot remains bottomed out at first, because damping element responds more slowly than an elastic element. If the stimulus continues long enough, a channel closes—the sign of adaptation—as the spring recoils and the piston extends in the dashpot. If a second larger stimulus is then applied, the spring stretches and the dashpot remains unmoved. Eventually, the spring would recoil and the dashpot's piston would extend to the end of its range, or top out, which would allow the channel to close, leading again to adaptation. This would explain adaptation, reactivation and a second bout of adaptation. When the stimulus is removed, a channel would recover, letting the spring and dashpot return to their original states.

Adaptation in *Xenopus* oocytes resembles that in hair cells. The major difference lies in their sensitivity to calcium ions: hair cells require an inward flow of calcium ions for adaptation, whereas mechanosensory adaptation in oocytes is calcium independent. In addition, *Xenopus* mechanosensory channels can switch from an adapting to a nonadapting mode if subjected to repeated, long pulses of suction. The channel current's adaptation decreases with successive pulses, and eventually the channel loses all mechanosensitivity. This behavior probably arises from a disruption of connections between the membrane and the cytoskeleton because of mechanical stresses from patch recording and stimulation. In fact, photomicrographs show that gentle suction on a patch leaves the membrane in close proximity to the underlying cytoplasm, and that maintained suction leads to a distinct separation between the membrane and the dark-pigmented

cytoplasm, presumably delineating the cytoskeleton of an oocyte. Although this cellular behavior represents artifacts related specifically to patch-clamp recording, the possibility exists than in different cells the presence or absence of membrane-cytoskeleton interactions may be programmed genetically to provide either adapting or nonadapting responses. Furthermore, such membrane-cytoskeletal interactions may be subject to modulation in a cell either during development or as a result of pathological conditions.

Models of Mechanosensitivity

Direct mechanical gating of a mechanosensory channel has two fundamental requirements. First, the channel needs a mechanism by which tension can be applied specifically to a channel molecule. Second, the channel molecule must be sensitive to the applied tension. Recent evidence indicates that tension may be exerted on a channel by a variety of mechanisms.

In bacteria, for example, a mechanosensitive-channel protein has been purified, cloned and sequenced. When this channel is reconstituted in cytoskeleton-free liposomes, essentially membrane-bound bags, it forms mechanosensitive channels. So tension is most likely exerted on the channel by elastic devices in the membrane itself. The reconstituted channels, though, display relatively low mechanosensitivity, and they do not show adaptation to sustained stimulation, suggesting that interactions with viscoelastic elements in the extracellular or cytoskeletal domains may be required to confer high sensitivity and adaptive behavior on mechanosensitive channels.

The concerted involvement in mechanoelectrical transduction of all three domains—extracellular membrane and cytoskeletal—is perhaps best exemplified by the specialized vertebrate hair cell. A hair cell's upper surface includes an array of slender hairs called a hair bundle. One "hair" is called the kinocilium, which is built like a true cilium, and the others are called stereocilia, which are built like the microvilli of the intestinal wall. In this system, extracellular "tip links," or fine filaments, interconnect individual stereocilia of the hair bundle. When a bundle in deflected, the top links deliver, or focus, tension on mechanosensitive channels in the membrane of the stereocilia. Furthermore, hair-cell adaptation supposedly involves an intracellular myosin-actin motor that actively resets the tip-link tension after adaptation, essentially taking up the slack, which poises membrane channels on the threshold of opening. Therefore, this motor is the basis for the exquisite sensitivity and adaptive behavior of a hair cell's mechanosensitive channels.

There is no information on the structure of mechanosensitive channels in oocytes and skeletal muscles or on how they are activated. Nevertheless, we have a speculative scheme for the elements behind mechanoelectrical transduction in these channels. To begin with, we assume that any mechanical stimulus applied to a cell—whether it be an overall stretching or localized perturbation—will distort the extracellular matrix as well as the membrane-cytoskeleton complex. Molecules that span the membrane may form part of a chain that focuses external mechanical energy on specific membrane-channel molecules. For example, proteins called integrins connect extracellular matrix molecules to cytoskeletal proteins. A variety of cytoskeletal proteins may also be links in this chain. In particular, proteins of the dystrophin-spectrin family are attractive candidates, because they lie directly beneath the membrane and have been shown to interact with a number of membrane glycoproteins. Interestingly, dystrophin has actin-binding domains that may generate active tension in the cytoskeleton, possibly priming mechanosensitive channels through a mechanism similar to the motor process of hair cells. Over the next few years, investigators should identify the exact molecular elements that are involved in this proposed chain of events

Bibliography

Brownell, P., and R.D. Farley. 1979. Detection of vibrations in sand by tarsal sense organs of the nocturnal scorpion. *Paruroctonus mesaensis*. *Journal of Comparative Physiology* **131**:23-30.

French, A.S. 1992. Mechanotransduction. *Annual Review of Physiology* **54**:135-152.

Guharay, F., and F. Sachs. 1984. Stretch-activated single ion channel currents in tissue-cultured embryonic skeletal muscle. *Journal of Physiology* **352**:685-701.

Hamill, O.P. 1983. Potassium and chloride channels in red blood cells. In *Single-Channel Recording*, ed. B. Sakmann and E. Neher. New York: Plenum Press.

Hamill, O.P., A. Marty, E. Neher, B. Sakkmann and F.J. Sigworth. 1981. Improved patch clamp techniques for high-resolution recording from cells and cell-free membrane patches. *Pflügers Archiv* **391**:85-100.

Hamill, O.P., J.W. Lane and D.W. McBride Jr. 1992. Amiloride: a molecular probe for mechanosensitive channels. *Trends in Pharmacology Sciences* **13**:373-376.

Hamill, O.P. , and D.W. Mcbride Jr. 1994. Molecular mechanisms fo mechanoreceptor adaptation. News in Physiological Sciences **9**: 53–59.

Hudspeth, A.J. 1989. How the ear's works work. *Nature* **341**:397-404.

Hudspeth, A.J., and P.G. Gillespie. 1994. Pulling springs to turn transduction: adaptation by hair cells. *Neurons* **12**:1-9.

McBride, D.W. Jr., and O.P. Hamill. 1992. Pressure-clamp: a method for rapid step perturbation of mechanosensitive channels. *Pflügers Archiv* **421**:606-612.

Morris, C. 1990. Mechanosensitive ion channels. *Journals of Membrane Biology* **113**:93-107.

Neher, E., and B. Sakkmann. 1992. The patch clamp technique. *Scientific American* March: 44-51.

Sukharev, S.I., B. Martinac, V.Y. Arshavsky and C. Kung. 1993. Two types of mechanosensitive channels in the *Escherichia coli* cell envelope: solubilization and functional reconstruction. *Biophysical Journal* **65**:177-183.

Wange, N., J.P. Butler and D.E. Ingber. 1993. Mechanosensitive across the cell surface and through the cytoskeleton. *Science* **260**:1125-1127. ❏

Questions

1. What are the two types of mechanoreceptors?

2. How do mechanoreceptors respond to touch?

3. How do pharmacological agents help in the study of ion channel function?

Answers are at the back of the book.

27

Sensations of smell are difficult to describe and classify, but useful categorizations have been made by noting the chemical elements of odorous substances. Research has pointed to the existence of seven primary odors—camphorlike, musky, floral, peppermintlike, ethereal (dry-cleaning fluid, for example), pungent (vinegarlike), and putrid—corresponding to the seven types of smell receptors in the olfactory-cell hairs. Odor receptors are responsible for correctly classifying scents into their proper categories. Neurobiologists believe that these receptors dissect our odorous world into distinct components and direct each component into a separate neural file. The treacherous journey of the odor receptor's message through the nasal passage to the olfactory cortex in the brain is now being studied by using molecular probes that tag olfactory receptors.

The Smell Files

Sarah Richardson

Discover, August 1995

Scent, the most primal of all senses, is arguably the most complex. The human nose contains millions of odor receptors of a thousand kinds—far more variety than is required for color vision or taste. How does the brain make sense of this flurry of signals? An essential first step, two groups of neurobiologists have recently found, is an orderly filing system. As they enter the brain from the different receptors, the hundreds of signals, each one representing a distinct odor component, are sorted into little round files called glomeruli—one type of signal per file.

Odor receptors lie in the mucous membranes high in the nose, on hairlike cilia that are the tips of dendrites—the receiving ends of the olfactory neurons. An individual neuron carries many copies of just one kind of receptor. When a smell molecule binds to a receptor, the receptor sends a signal up the dendrite to the cell body. From there it travels down the neuron's sending arm, the axon. The axon passes through a hole in the bone into one of the olfactory bulbs—twin structures, each roughly the size of a small grape, that are lodged above each nostril on the underside of the brain.

The glomeruli (the word is Latin for "little balls") line the olfactory bulbs. Each one is a neural junction, a place where the axons from roughly 2,000 olfactory neurons meet and pass signals to the dendrites of some 30 mitral cells. The mitral cells then refine the signals and relay them to the olfactory cortex, a higher region in the brain.

That much has been known for some time now. And so it has also been clear for some time that glomeruli must play a crucial role in organizing scent perception. But what hasn't been clear, because the neural connections hadn't been mapped out, is exactly what that role is. Past studies had shown that different odors activate different subsets of glomeruli; and yet a single glomerulus was also thought to respond to a variety of odors. Is it receiving information from neurons bearing different receptors, or does it receive information from only one kind of olfactory receptor?

By using molecular probes that tag particular olfactory receptors, researchers have recently untangled some of the neural wiring in the olfactory bulbs of rats and shown which wire connects to which junction. Both groups—one led by Richard Axel at the Howard Hughes Medical Institute at the Columbia College of Physicians and Surgeons and the other by Linda Buck of Harvard Medical School —found that glomeruli don't get mixed signals. Instead, each gets a very clear one: only neurons bearing one kind of receptor converge on a glomerulus. In fact, it looks as though there are roughly as many glomeruli as there are kinds of olfactory receptors. A glomerulus for a particular receptor seems to be located in the same place in each of the two olfactory bulbs, and in the same place in all rats.

Each receptor and each glomerulus, Axel explains, responds to just one part of a smell molecule's structure rather than the whole molecule. "What you're doing is essentially dissecting the odorous image," he says. "You deconstruct the image such that the individual components of a given odor— even an odor elicited by a single molecule—will react with different receptors simultaneously." Because different molecules may share some structural features, a single glomerulus may be activated by different odors, even though it receives information from only one kind of receptor.

Any given odor, however, is distinguished by the pattern of glomeruli it activates. Based on this molecular fingerprint, the brain can somehow recognize nearly 10,000 scents, despite having a repertoire of only a thousand olfactory receptors. Glomeruli help us do that not only by sorting the signals but also by enhancing sensitivity, says Buck. The convergence of similar neurons in a single glomerulus helps the brain get enough of a sample to recognize the odor, even if it is present in very small quantities.

This method of identifying odor components may also explain why we can recognize odors we haven't smelled in decades. Without stimulation, says Buck, olfactory neurons will die. Since an individual neuron responds to a component that many odors share rather than to one particular odor, it is stimulated often. That may keep our neurons in fighting trim, Buck speculates, and able to recognize a distinctive pattern of a scent long after it was first encountered.

The most tantalizing question of all, however, remains unanswered. How is the map of activated glomeruli decoded in the brain? The precise positioning of glomeruli within the olfactory bulb probably simplifies the brain's job somewhat. "If the position to which an olfactory neuron projects is fixed," says Axel, "then the brain can use anatomic position as an indicator of odor quality." But to find out if that's what is really going on, he and Buck will have to untangle the nerve knots some more. "The next question is to go one step further into the cortex," says Axel, "and ask how the mitral cells project to the cortex. That might give us some indication of how this map is read." ❏

Questions

1. What are the names of the receiving ends of the olfactory neurons?

2. What occurs at the glomeruli?

3. What distinguishes one odor from another?

Answers are at the back of the book.

28 *Nothing is more satisfying to the senses than the tastes of a fine meal. The tastants responsible for simulating our senses enter at the taste pore, the opening of the taste bud, and bombard the taste-receptor cells within. Neurons make contact with the taste-receptor cells at the synapse where signals are passed via chemical transmitters and are then simply forwarded to the brain. Sounds simple, but our sense of taste, however, demonstrates more biological complexity when viewed at the molecular level. This is especially true in the sweet-taste and bitter-taste transduction pathways. Interaction of these tastants with receptors activates a diverse family of taste cells called G proteins, for GTP-binding proteins. Of interest to scientists is the new G protein isolated from taste cells called gustducin, which is specifically expressed there. Gustducin may be involved in the transduction of bitter tastes through to mechanism similar to that used in visual transduction. This draws an interesting parallel between the two senses and provides many new questions regarding the molecular mechanisms involving taste.*

The Sense of Taste

Susan McLaughlin and Robert F. Margolskee

American Scientist, November/December 1994

The internal molecular workings of the taste bud help it distinguish the bitter from the sweet.

To the epicure, nothing represents the height of refinement better than the diverse tastes of a fine meal. To the biologists, nothing could be more basic than the molecular mechanisms underlying taste. The chemical senses—taste and smell—are the most primitive of the sensory modalities. Before warning labels were invented, people had to rely on their sense of taste to distinguish nutritionally useful substances, which taste good, from potentially toxic substances, which taste bad. Today, people still rely on their sense of taste to avoid spoiled foods and fine-tune their diets to specific nutritional needs of the moment. A body that has just vigorously exer-

cised may crave salty food; one low on sugar will yearn for something sweet.

Surprisingly, the many taste sensations that delight the epicure are produced from only four types of tastes: salt, sour, sweet and bitter. Like primary colors, these tastes can be blended and combined to create the many shades and hues of flavor. In addition, the sense of small contributes greatly to the totality of flavor perception. Chewing releases volatile chemicals into the nasal passages; the scents of these chemicals combine with tastes to create even more flavors. That is why that sinfully delicious piece of chocolate cake is so much less delectable when you nose is stuffed during a head cold.

At the molecular level, taste is quite similar to our other senses. A gustatory stimulus activates a taste-receptor cell, which in turn conveys the message to a

Reprinted with permission from *American Scientist*, Vol. 82, November/December 1994, pp. 538–545.

sensory neuron. A nerve impulse then relays the message to the gustatory centers of the brain, where it registers as a taste. Because taste stimuli are restricted to only four classes, it might be assumed that conveying information about taste would be a simple task. If one were to design a taste system, the obvious solution would be to create four distinct receptor cells, one for each taste category. Each type of receptor cell would then uniquely respond to and transmit information about its own class of stimulus (the so-called "labeled-line" model).

Unfortunately, nature and evolution have not chosen this simple solution. Instead, it appears that each taste receptor cell is capable of responding to more than one class of stimulus. Therefore, the specificity in the system must occur elsewhere. If a single taste receptor cell can respond to several types of stimuli, then it might achieve specificity by treating each type of stimulus differently. Indeed, the picture emerging from molecular studies shows that a receptor cell uses a separate processing pathway for each class of taste stimulus.

Increasing the complexity of the situation is the fact that stimuli for a single taste modality may themselves come from several different types of chemicals. Bitter tastes, for example, can be elicited by compounds as diverse as caffeine, which is chemically classified as a purine; morphine, and alkaloid; or potassium chloride, a simple salt. Not only does each separate taste modality require its particular coded pathway, but each taste modality may use more than one processing mechanism.

The mechanism of taste perception has been of interest for thousands of years. Over 2,000 years ago, the Roman philosopher and poet Lucretius suggested "that formed of bodies round and smooth are things which touch the senses sweetly, while all those which harsh and bitter do appear, are held together bound with particles more hooked, and for this cause are wont to tear their way into our senses, and on entering in to rend the body." Although this description of taste would not be well received in scientific circles today, it nevertheless contains a fundamental truth. The shape of a molecule that stimulate taste does indeed determine which taste modality it stimulates.

The scientific understanding of the way that taste cells process different stimuli has certainly progressed since Lucretius's time, but many questions remain. The surprising complexity of what would at first glance appear to be a simple process has stimulated the efforts of biochemists, physiologists and psychophysicists for many years.

Taste Buds

Found primarily on the tongue, taste cells (also called taste-receptor cells) are organized into specialized structures called taste buds—gourd-shaped collections of 50-150 taste cells. Taste-receptor cells are long and spindle-shaped with protrusions at their tips called microvilli. The first step in taste recognition takes place at the taste pore, an opening in the taste bud where the microvilli of the receptor cells contact the outside environment. Molecules that are perceived to have taste, called tastants, enter the taste pore and interact with receptor molecules and channels within the microvillar membrane of the taste-receptor cells.

Microscopic examination of taste buds reveals several distinct cell types; light cells, dark cells, intermediate cells and basal cells. The specific function of each cell type is not yet known. Each may represent a different stage of taste-cell maturation. Basal cells mature into dark cells, which then become light cells.

The life span of an individual taste cell is only 10 days to 2 weeks, so cells within the taste bud are continually being replaced. As a consequence, cells at many different stages of development are present within a single taste bud.

Mammalian taste buds are localized in structures call papillae (Latin for "bumps"). The numbers and types of papillae vary depending on the species. Rats and people, for example, have three types of taste papillae: fungiform, foliate and vallate. Each type of papilla is found on a different area of the tongue. Fungiform papillae are relatively simple structures, generally containing only single taste bud. Several hundred fungiform papillae are scattered randomly near the tip of the rate tongue. Foliate and vallate papillae are more complex structures, each containing many taste buds. Foliate papillae consist of a series of parallel grooves along the sides of the tongue. Taste buds line the sides of the grooves, with their pores facing the cleft. Taste buds also line the clefts of the vallate papillae, which are circular or horseshoe-shaped trenches at the back of the tongue.

Besides detecting taste stimuli, taste-receptor cells must also convey taste information to the brain. The link between the taste-receptor cell and the brain is the neuron. Neurons infiltrate taste buds and make contact with taste cells at the synapse, a specialized region between the receiving end of the neuron and the sending end of the taste cell. Information is passed from the taste-receptor cell to the neuron by means of chemical transmitters, called neurotransmitters, secreted by the taste cell into the synapse. Neurons detect these transmitters and react to them with a nerve impulse that is transmitted to the brain. The act of receiving sensory information and then translating it into a signal useful to the nervous system is called sensory transduction.

Salt

Electrically charged ions can pass into and out of cells through specialized protein pores, call ion channels, which are embedded within cell membranes. The quintessential salty taste is the positively charged sodium ion ($Na+$) found in common table salt, sodium chloride. Prior to 1981, it was thought that taste cells were impermeable to ions; however, in that year, John DeSimone, Gerard Heck and Shirley DeSimone of the Medical College of Virginia showed that sodium ions could be actively transported across epithelial membranes isolated from dog tongue. They also showed that sodium-ion transport across the membrane could be inhibited by the drug amiloride, which blocks sodium-ion transport through specialized sodium channels. These results suggested that taste-cells contain sodium-ion channels. Further studies in 1983 by Susan Schiffman, Elaine Lockhead and Frans Maes of Duke University showed that in humans and rats, the salty perception of sodium chloride is inhibited by amiloride. John DeSimone and his coworkers proposed that sodium ions enter taste cells by passive diffusion through amiloride-sensitive sodium channels.

The interior of an unstimulated taste cells is nega-

tively charged with respect to the external environment. When sodium ions enter the taste cell, the net charge on the inside of the cell becomes less negative—that is, the charge difference between the cell's interior and exterior become less polarized. This depolarization causes the taste cell to release neurotransmitters into the synapse between the taste cell and the adjacent neuron. The nerve responds to the neurotransmitters and relays the information to the brain.

There may be other mechanisms for salt-taste transduction. In some species between 30 and 50 percent of the response to sodium chloride is not blocked by amiloride.

Sour

Whereas many salts taste salty, acids taste sour. Sour-tasting acids can be inorganic compounds, such as hydrochloric acid, or organic, such as lactic acid, but they all share a common characteristic: They all release a positively charge hydrogen ion ($H+$). The hydrogen ions released by acids produce the sour taste.

Sue Kinnamon and Stephen Roper of Colorado State University showed that weak acids stimulate mudpuppy (the amphibian *Necturus*) taste cells, and that this stimulation could be inhibited by chemicals that block potassium ion ($K+$) channels. This suggested that potassium channels were involved in sour-taste transduction. The proposed mechanism for sour-taste in amphibians is that hydrogen ions block potassium channels, preventing the release of positively charged potassium ions from the taste cell, leading to cell depolarization and subsequent neurotransmitter release.

However, in hamsters, sour-taste transduction may operate through a different mechanism. Timothy Gilbetson, Patrick Avenet, Sue Kinnamon and Stephen Roper of Colorado State University showed that the hydrogen ions can center the taste cell through sodium-ion channels and directly cause cell depolarization.

Sweet

The molecular mechanisms underlying salt- and sour-taste transduction involve a direct interaction between the stimuli, which are ions, and the ion channels. However, the mechanisms for the detection of sweet and bitter compounds are more complex.

Sweet stimuli are large and chemically more complex than ions, and cannot enter taste cells through ion channels; rather, sweet stimuli contact specialized receptor molecules on the taste cell's surface. The contact between the sweet stimulus and a surface receptor molecule triggers a flurry of chemical signals inside the taste cell that culminates in the release of neurotransmitter.

The internal chemical signals generated in response to external stimuli are called second messengers. Second messengers include ions (frequently the calcium ion), large fatty molecules (for example, the lipids phosphatidyl inositol and diacylgycerol), and small molecules (for example, the cyclic nucleotides cAMP and cGMP). Second-messenger pathways involve a long string of chemical reactions where the product of one reaction is the activator for the next, until finally the last event in the chain stimulates a particular cellular activity or cellular output.

Particularly important regulators of second messengers are the GTP-binding proteins, or G proteins. There are two main types of G proteins: small G proteins (for example, Ras, a G protein implicated in carcinogenesis), and the large heterotrimeric G proteins. Heterotrimeric G proteins make up a diverse family of proteins involved in a signal transduction in many different cell types, including smooth-muscle cells, neurons and sensory cells.

The heterotrimeric G proteins are made up of three subunits; alpha, beta and gamma. The specificity of the G protein is largely dependent on the alpha subunit, which interacts with the cytoplasmic face of a specific receptor molecule embedded within the cell's membrane. The alpha subunit ties the G protein to the next component in the transduction pathway. In effect, the alpha subunit determines in which chemical pathway the G protein lies.

Different G-proten alpha subunits activate different second-messenger pathways in the cell. Stimulatory G proteins (G_s and G_{olf}) stimulate the production of the cyclic nucleotide cAMP by acivating the enzyme adenylyl cyclase, which makes cAMP from ATP. Inhibitory G proteins (G_{i1}, G_{i2}, and G_{i3}) inhibit adendylyl cyclase and thus block the formation of cyclic nucleotides.

G proteins and cyclic nucleotides play an important role in the first steps in the visual-transduction pathway. The rod and cone cells of the retina contain G proteins call transducins ($G_{t\text{-rod}}$ and $G_{t\text{-cone}}$). Transducins activate the enzyme phosphodiesterase, which converts cGMP into GMP.

G proteins of the G_q class (G_q and G_{14}) activate the enzyme phospholipase C to generate inositol trisphosphate (IP_3) and diaclyglycerol (DAG), which serve as second messengers. Other G proteins, such as G_o, interact directly with ion channels.

Although they are involved in many different second-messenger pathways, all G proteins are activated in a similar fashion. When the appropriate molecule binds to the exterior face of cell-surface receptor, the receptor undergoes a conformational change into its active state; the activated receptor in turn activates the G protein inside the cell. The subunits of the activated G protein then dissociate, and the alpha subunit interacts with the next component in the pathway. Eventually, the alpha subunit reassociates with the other two subunits, and the G-protein complex awaits the next signal from the cell surface receptor.

Beginning in 1988, evidence began to accumulate that suggested the participation of G protein in the transduction of sweet tastes. Keiichi Tonosaki and Masaya Funakoshi of Asahi University showed that sucrose caused the depolarization of taste cells. Depolarization was also seen when cGMP or cAMP was injected into the cells, suggesting that the transduction of sucrose involved the formation of cyclic nucleotides. In the same year, Patrick Avenet, F. Hoffman and Bernd Lindemann of Saarlandes University showed that the injection of cAMP into frog taste cells resulted in cell depolarization owing to decrease in the flow of potassium ions out of the cell. In addition, they showed that high levels of cAMP in the frog taste cell activated a protein kinase, an enzyme that induces the closure of potassium-ion channels. The enzyme adenylyl cyclase is present at high levels in taste cells and in 1989, Benjamin

Striem, Umberto Pace, Uri Zehavi and Michael Naim of Hebrew University and Doron Lancet of the Weizmann Institute showed that stimulation of adenylyl cyclase by sweet substances requires guanine nucleotides, suggesting that a G protein (presumably of the Gs family) is involved.

These experiments suggested a possible mechanism for sweet-taste transduction. When a sweet tastant binds to a receptor on the cell surface, a stimulatory G protein (G_s) is activated inside the taste cell. The alpha subunit of G_s dissociates from the G protein complex and stimulates adenylyl cyclase to produce cAMP. The cAMP then activates a protein kinase, which chemically modifies and closes the potassium channel. Closing the potassium channel blocks the exit of positively charged potassium ions, causing the taste cell to depolarize and release neurotransmitter.

Although some of the intracellular molecules that transduce sweet taste have been identified, the nature of the extracellular sweet receptor is currently unknown. Sweet-tasting compounds are chemically very diverse, and include polysaccharides as well as amino acids, proteins and various artificial compounds, such as saccharin and aspartame. It seems unlikely that a single receptor molecule can accommodate all sweet stimuli.

Bitter

Bitter-taste stimuli are the most chemically diverse of all tastants. Some bitter substances have affinity for the fatty lipid molecules in the cell membrane—these substances are termed lipophilic ("lipid-loving"). Lipophilic bitter substances may penetrate the membrane directly, whereas hydrophilic ("water-loving") bitter tastants may interact with a cell-surface receptor. In addition, experimental evidence suggests that there may be more than one intracellular signaling pathway involved in bitter transduction.

The detection of bitter substances is extremely important for the survival of an organism; sensitivity to bitter stimuli is a protective mechanism for poison avoidance. Many plant and environmental toxins taste bitter. This provides a strong evolutionary pressure to be highly sensitive to a bitterness. The diversity of mechanisms for detecting bitterness may lead the ability to detect many different kinds of bitter compounds.

There is strong experimental evidence for one mechanism of bitter-taste transduction. Studies done in 1988 by Myles Akabas, Jane Dodd and Qais Al-Awaqati of Columbia University, and in 1990 by Solomon Snyder and coworkers at the Johns Hopkins University have helped to establish the participation of IP_3, and DAG, and implicate the participation of a G_q protein in bitter taste transduction.

Their experiments suggest that some bitter molecules bind to a cell-surface receptor that activates G_q. Activated G_q activates a phospholipase C to genrate IP_3, which is known to stimulate the release of calcuim ions from internal cellular stores. Calcium ions can cause cell hyperpolarization by activation of potassium channels; calcium can also directly activate neurotransmitter release.

Evidence of other bitter transduction pathways is controversial. Studies have linked bitter substances to phosphodiesterase activity. Some experiments, however, suggest that bitter tastants activate phosphodiesterase, while others suggest that bitter substances inhibit phosphodiesterase.

Still a third mechanism has recently been proposed by Yukio Okada, Takenori Miyamoto and Toshihide Sato of Nagasaki University, who have linked bitter substances with chloride ion secretion, which leads to cell depolarization.

Unique Elements

Many of the molecules involved in taste transduction are common players in the signal-transduction pathways of other cell types. However, taste cells are specialized to detect tastants, so it would not be surprising to find that they use a preferred set of signaling molecules, some of which might be unique to taste cells. In our laboratory, we used molecular biological techniques to identify and clone specific components of taste-transduction pathways. These techniques allowed us to isolate the genes encoding some of the components of taste-transduction pathways.

Our approach has been particularly successful in identifying the G proteins expressed in taste cells.

Using a technique called the polymerase chain reaction, we identified three known G proteins expressed more highly in taste cells than in non-taste tongue (epithelium G_{13}, G_{14} and G_s). Since these G proteins are expressed in high levels in taste cells, they may be involved in taste transduction.

G_{13} is a member of the inhibitory class of G proteins, which prevent adenylyl cyclase from making cAMP. G_{14} is a G_q-type protein, which leads the generation of the second messenger IP_3. One of the mechanisms for bitter taste transduction implicates a G_q-like protein and IP_3 generation, suggesting that G_{14} might be involved in bitter-taste transduction. The stimulatory G protein G_s stimulates adenylyl cyclase, leading to an increase in cAMP.

In addition to these previously characterized proteins, we isolated a novel taste-specific G-protein alpha subunit. Surprisingly, the inferred amino acid sequence of this new protein most closely resembled that of the transducins, the G proteins involved in phototransduction. We therefore named this new G protein the alpha subunit gustducin, for gustatory transducin. In subsequent experiments we showed that gusducin is expressed specifically in taste cells.

The close relationship of gustducin to the transducins came as a complete surprise. Based on previous experimental evidence, one might expect that the G proteins involved in taste transduction would be G_s (for sweet taste transduction) or G_q (for bitter taste transduction). The isolation of a taste-specific G protein that resembled the retina-specific transduction suggested that the transduction of chemical taste stimuli might, at least in some ways, be similar to the transduction of light. This is certainly not an intuitive expectation. What role does gustducin play in taste cells? Its close relationship to the tranducins provides us with some clues and leads us to a testable hypotheses.

To understand the role of gustducin in taste transduction, it may be useful to examine how the visual transducins transduce light stimuli. Light enters the rod and cone photoreceptor cells of the retina, and activates light-sensitive receptor (rhodopsin) molecules. Activated rhodopsin then activates transducin. The alpha subunit of transducin stimulates phosphodieterase, which degrades the second messenger molecule cGMP. Decreased levels of cGMP in the rod and cone cells cause positive-ion channels in the cell's membranes to close.

We can make some predictions about how gustducin might operate by comparing its structure to that of the transducins. The regions of the transducin protein that interact with rhodopsin and phosphodiesterase closely resemble the same regions in the gustducin protein. This suggests that gustducin interacts with a membrane receptor in taste cells that may be similar the visual receptor rhodopsin. In addition, we would predict that gustducin also interacts with a phosphodiesterase in taste cells.

In which type of taste transduction might gustducin be involved? If gustducin activates a phophodiesterase, then we would expect the levels of cyclic nucleotides (cAMP or cGMP) inside the taste cell to fall following gustducin activation. Since sweet substances are known to increase the amounts of cAMP in taste cells, it is unlikely that gustducin is involved in the detection of sweet tastes.

A more plausible model implicated gustducin in bitter taste transduction. We propose that bitter substances interact with receptors on the taste-cell surface, which in turn activates gustducin. Activated gustducin would then stimulate a phosphodiesterase to degrade cyclic nucleotides. Specialized channels open and allow the passage of sodium, potassium and calcium ions into the cell when cyclic nucleotide levels are low. The entrance of these positively charged ions depolarizes the taste cell, and neurotransmitter is release.

If bitter taste is transduced by mechanisms involving gustducin and phosphodiesterase, or G_q and phospholipase C, how is sweet taste transduced? Activation of G_s proteins leads to an increase in cAMP, which is observed in taste cells exposed to sweet stimuli. Therefore, G_s is the most likely candidate for mediating sweet-taste transduction, and this in consistent with our finding that G_s expression is elevated in taste cells.

A Taste of the Future

To date, the G protein gustducin is the only taste cell-specific protein that has been identified. Taste-transduction pathways may contain other components that are taste-cell specific (for example, taste receptor molecules), or they may be made up of proteins that also transduce signals in other cell types.

Certainly the most interesting discovery is the close relationship of the taste-specific gustducin to the retina-specific tranducins. It suggests that there is a previously unsuspected level of evolutionary conservation at the molecular level between taste transduction and phototransduction. Perhaps taste-cell receptor molecules resemble rhodopsin. The parallels between taste and vision suggest future experiments that would involve mixing and matching taste and visual transduction components to determine whether parts of these two processes are interchangeable.

Many questions about taste transduction remain to be answered, and molecular biology may provide further insight into taste transduction pathways. In all, studies on taste transduction promise to be very sweet for many years to come.

Bibliography

Akabas, M.H., J. Dodd and Q. Al–Awqati. 1988. A bitter substance induces a rise in intracellular calcium in subpopulation of rat taste cells. *Science* **242**:1047–1050.

Avenet, P., and B. Lindemann. 1989. Perspectives of taste reception. *Journal of Membrane Biology* **112**: 1–8.

Avenet, P., F. Hoffmann and B. Lindemann. 1988. Transduction in taste receptor cells requires cAMP-dependent protein kinase. *Nature* **331**:351–354.

Behe, P., J.A. DeSimone, P. Avenet and B. Lindemann. 1990. Membrane currents in taste cells of the rat fungiform papilla. *Journal of General Physiology* **96**:1061–1084.

Derma. 1947. *Proceedings of the Oklahoma Academy of Science* **27**:9

DeSimone, J.A., G.L. Heck and S.K. DeSimone. 1981. Active ion transport in dog tongue: A possible role in taste. *Science* **21**:1039–1041.

DuBois, G.E., D.E. Walters, S.S. Schiffman, Z.S. Warwick, B.J. Booth, S.D. Pecore, K. Gibes,

B.T. Carr and L.M. Brands. 1991. Concentration–response relationships of sweeteners: A systematic study. In *Sweeteners: Discovery, Molecular Design, and Chemoreception*, eds. D.E. Walters, F.T. Ortheofer and G.E. DuBios, Washington, D.C.: American Chemical Society, p. 261

Frank, M., and C. Pfaffmann. 1969. Taste nerve fibers: A random distribution of sensitivities to four tastes. *Science* l **164**:1183–1185.

Gilbertson, T.A., P. Avenet, S. C. Kinnamon and S.D. Roper. 1992. Proton currents through amiloride-sensitive Na channels in hamster taste cells: role in acid transduction. *Journal of General Physiology* **100**:803–824.

Gilbertson, T.A. 1993. The physiology of vertebrate taste reception. *Current Opinion in Neurobiology* **3**:532–539.

Heck, G.L., S. Mierson and J.A. DeSimone. 1984. Salt taste transduction occurs through an amiloride–sensitive sodium transport pathway. *Science* **223**:403–405.

Hwang, P.M., A. Verma, D.S. Bredt and S. Snyder, 1990. Localization of phosphatidylinositol signaling components in rat taste cells: role in bitter taste transduction. *Proceedings of the National Academy of Sciences* **87**:7395–7399.

Kinnamon, S.C. 1988. Taste transduction: a diversity of mechanisms. *Trends in Neuroscience* **11**:491–496.

Kinnamon, S.C., and S.D. Roper. 1988. Membrane properties of isolated mudpuppy taste cells. *Journal of General Physiology* **91**:357–371.

Kinnamon, S.C., V.E. Dionne and K.G. Beam. 1988. Apical localization of K channels in taste cells provides the basis for sour taste transduction. *Proceedings of the National Academy of Sciences* **85**:7023–7027.

Margolskee, R.F. 1993. The biochemistry and molecular biology of taste transduction. *Current Opinion in Nerurobiology* **3**:526–531.

Margolskee, R.F. 1993. The molecular biology of taste transduction. *BioEssays* **15**:645–650.

Pfaffman. 1959. *Handbook of Physiology*, Sec. I., Vol.

I. Baltimore: The Williams and Wilkins Co., p. 507.

Roper, S.D. 1983. Regenerative impulses in taste cells. *Science* **220**:1311–1312.

S.K. McLaughlin, P.J. McKinnon and R.F. Margolskee, 1992. Gustducin is a taste-cell specific G Protein closely related to the transducins. *Nature* 357:563–569.

Schiffman, S.S., E. Lockhead and F.W. Maes. 1983. Amioloride reduces the taste intensity of Na$^+$ and Li$^+$ salts and sweeteners. *Proceedings of the National Academy of Sciences* **80**:6136–6140.

Striem, B.J., M. Naim and B. Lindemann. 1991. Generation of cyclic AMP in taste buds of the rat circumvallate papillae in response to sucrose. *Cellular Physiology and Biochemistry* **1**:46–54.

Striem, B.J., U. Pace, U. Zehavi, M. Naim and D. Lancet. 1989. Sweet tastants stimulate adenylate cyclase coupled to GTP binding protein in rat tongue membranes. *Biochemistry Journal* **260**:121–126.

Tinti, J.M., and C. Nofre. 1991. Why does a sweetener taste sweet? A new model. In *Sweeteners: Discovery, Molecular Design, and Chemoreception*, ed. D.E. Walters, F.T. Orthoefer and G.E. Dubois. Washington, D.C.: American Chemical Society, pp. 206–213.

Tonosaki, K., and M. Funakoshi 1988. Cyclic nucleotides may mediate taste transduction. *Nature* **331**:354–356. ❑

Questions

1. What are the four types of tastes?

2. What is the location of the taste-receptor?

3. What second messengers are involved in the signal transduction of sweet and bitter compounds?

Answers are at the back of the book.

29   *Very little is known about the molecules and pathways that mediate the sensory signals of taste. Although we higher organisms only exhibit four types of taste modalities, salty, sour, sweet and bitter, a myriad of signaling pathways must create the proper strategy for taste transduction to be successful. The tastes of sour and salt are believed to modulated by the entry of H+ and Na+ ions through specialized channels of the membrane while sweet and bitter transduction is coupled with G-protein activity. Interestingly, enough, a molecule in the bitter signaling cascade called gustducin, has been found to be very similar to the visual transducin used for coupling the light-receptor molecule rhadopsin. Using this information, scientists hope to probe the area of taste transduction and gain insight into the biology of signaling pathways.*

A Taste of Things to Come

Charles S. Zuker

Nature, July 6, 1995

Taste transduction is one of the most sophisticated forms of chemotransduction,[1,2] operating throughout the animal kingdom from simple metazoans to the most complex of vertebrates. Its purpose is to provide a signalling response to non-volatile ligands (olfaction provides a highly specific, extremely sensitive response to volatile ligands). Higher organisms have four basic types of taste modality: salty, sour, sweet, and bitter. Each of these is though to be mediated by distinct signaling pathways that lead to receptor cell depolarization, release of neurotransmitter and synaptic activity.[3] Although much is known about the psychophysics and physiology of taste cell function,[4] very little is known about the molecules and pathways that mediate these sensory signaling responses. To complicate matters, some taste-receptor cells respond to more than one taste modality, raising important mechanistic questions about signal crosstalk and information encoding and decoding. Now[5,6] Margolskee and colleagues offer us some surprising insights into these signaling cascades.

Sour and salty tastants are believed to modulate taste cell function by direct entry of H^+ and Na^+ ions through specialized membrane channels localized on the apical surface of the cell. In the case of sour compounds, taste-cell depolarization is thought to result from H^+ blockage of K^+ channels, and salt transduction from the entry of Na^+ through amiloride-sensitive Na^+ channels. None of the molecular components of the sour or salty pathway has be identified.

Sweet and bitter transduction are believed to be G-protein-coupled transduction pathways. However, nothing is known about the membrane receptors activated by bitter and sweet compounds—this is surprising, given the tremendous biotechnological and pharmaceutical applications for bitter antagonists and sweet agonists. The transduction to sweet tastants appears to involve cyclic AMP as a second messenger.[8-10] The current view is that a seven-helix transmembrane sweet receptor activates a G_s type G protein, which in turn activates adenylate cyclase. Increased cAMP leads to the activation of protein kinase A and the phosphorylation and blockage of a family of potassium channels. In this regard, cAMP has been shown to mimic the effect of sweeteners, leading to receptor-cell depolarization via a blockage of a potassium conductance. Interestingly, there is also evidence for an amiloride-sensitive conductance in sweet transduction.[11]

Bitter transduction appears to make use of a number of intracellular pathways relying on different signaling strategies. For instance, denatonium, one of the most bitter substances known to humans, leads to release of calcium from intracellular stores in some taste-receptors neurons.[12] This suggests an involvement of phosphoinositide-phospholipase C (PLC) signaling in this pathway. But cyclic nucleotide phosphodiesterases (PDE) have also been implicated in bitter transduction. The PLC pathway probably involves the modulation of Ca^{2+}-activated conductances, whereas the PDE pathway may alter the gating of cyclic-nucleotide-modulated ion channels.

The only molecule to be identified so far in the bitter signaling cascade is gustducin,[13] a $G\alpha$ subunit that is highly specific to taste neurons and very similar to visual transducin. Transducin is the $G\alpha$ subunit in vertebrate photoreceptor cells and is responsible for coupling the light-receptor molecule rhodopsin to a photo-receptor-cell-specific cGMP phosphodiesterase. Active PDE hydrolyses cGMP into GMP and because the light-regulated channels are opened, the transient reduction in cGMP causes the closing of these cation-selective channels. Thus, the absorption of a photon by the vertebrate photoreceptors causes a brief hyperpolarization of a cell.

In their papers,[5,6] Margolskee and colleagues report two intriguing findings on vertebrate taste receptor cells. The first, by Ruiz-Avila et al.,[5] demonstrates the presence of visual transducin in rat and bovine taste-receptor neurons. The authors describe studies involving immunohistochemistry and the polymerase chain reaction to show that transducin is expressed in some taste receptor cells; moreover, indirect assays provide hints that transducin functions in bitter taste transduction. These results are highly suggestive and will fuel research in this area, but one must be cautious about extrapolating expression studies into functional ones—particularly important here as gustducin and transducin share about 80 percent amino-acid identity. The most rigorous experiment, namely the generation of transducin-knockout mice, is yet to be done, although gustducin knockouts have recently been produced (R. Margolskee et al., unpublished results). The analysis of these animals as well as transducin/gustducin double-mutants should provide important insight into the biology of these pathways. Interestingly, if transducin does indeed function in coupling a bitter receptor to phosphodiesterase, then it follows that the receptor in question is likely to share significant homology with rhodopsin.

In the second paper,[6] Kolesnikov and Margolskee describe the presence of a cyclic-nucleotide-suppressible conductance in frog taste-receptor neurons. This is something of a novel mechanism, as cyclic nucleotides have never been shown to suppress or inactivate ion channels directly. This is in contrast to cyclic-nucleotidegated ion channels in which direct binding by cAMP or CGMP opens the channels. Single-channel studies should provide a detailed description of the biophysical properties of these channels and the role of cyclic nucleotides in their regulation. Kolesnikov and Margolskee also show that the electrophysiological responses of some receptor neurons to the two sweeteners saccharin and NC-01 (assuming that both taste sweet to frogs) are modulated by IBMX, an inhibitor of phosphodiesterase, suggesting a role for phosphodiesterase in their transduction pathway.

Given that most sweet transduction responses are believed to be mediated through activation of adenylate cyclase, the result presented in these two studies give us a taste of the complexity and excitement of this field.

References

1. Avenet, P. & Lindemann, B. *Persp. Taste Recep.* **112**, 1-8 (1989).
2. Margolskee, R. *Bioessays* **15**, 645-650 (1993).

3. Roper, S.D. *A. Rev. Neurosci.* **12**, 329-353 (1989).

4. Gilbertson, T. *Physiol. Vert. Taste recep.* **3**, 532-539

5. Ruiz-Avila, L. *et al. Nature* **376**, 80-85 (1995).

6. Kolesnikov, S.S. & Margolskee, R.F. *Nature* **376**, 85-88 (1995).

7. Kinnamon, S. *Trends Neurosci.* **11**, 491-496 (1988).

8. Avenet, P., Hoffman, F. & Lindemann, B. *Nature* **331**, 351-354 (1988)

9. Striem, B., Pace, U., Zehavi, U., Naim, M. & Lancet, D. *Biochem.* **260**, 121-126 (1989).

10. Striem, B., Niam, M. & Lindemann, B. *Cell Physiol. Biochem.* **1**, 46-54 (1991).

11. Schiffman, S., Lockhead, E. & Maes, F. *Proc. Natn. Acad. Si. U.S.A.* **80**, 6136-6140 (1983).

12. Akabas, M., Dodd, J., & Al-Awquti, Q. *Science* **242**, 1047-1050 (1988).

13. McLaughlin, S., McKinnon, P. & Margolskee, R. *Nature* **357**, 563-569 (1992). ❏

Questions

1. What are the four basic types of taste modalities?

2. How is a response in the receptor elicited by taste?

3. What is gustducin?

Answers are at the back of the book.

Section Five

Important Messengers of the Renal, Endocrine, Respiratory, Cardiovascular, and Reproductive Systems

30 *The list of hemoglobin's capabilities is fast expanding, due to the recent discovery that it can carry a form of the ever popular molecule nitric oxide. It all started when scientists began wondering what might bind the two cysteines on the hemoglobin molecule. Much to their surprise, hemoglobin reacted with S-nitrosothiols (SNO) in the lungs and later released it at crucial moments directly into the blood vessels. What's most exciting about the molecular marriage of SNO and hemoglobin is its potential to stabilize blood pressure. Using hemoglobin as a shuttle service for nitric oxide will open up new research in developing blood substitutes that are more blood pressure friendly.*

Hemoglobin Reveals New Role as Blood Pressure Regulator

James Glan

Science, March 22, 1996

Hemoglobin may have been leading a double life, right under biochemists' noses. This familiar constituent of red blood cells, which delivers oxygen to tissues and retrieves carbon dioxide, is probably the most studied protein in existence. But a group led by Jonathan Stamler at the Duke University Medical Center reports in yesterday's issue of *Nature* that, alongside the familiar respiratory cycle, hemoglobin carries out a second cycle in which it sops up a form of nitric oxide (NO) in the lungs and releases it in blood vessels—a shuttle service that helps stabilize blood pressure. "Just when we thought we knew everything about hemoglobin, we've discovered something new," says co-author Joseph Bonaventura. "It's gratifying and exciting."

The results connect two molecules, NO and hemoglobin, that had been seen as biochemical enemies. NO, produced in the endothelial cells that line the blood vessels, relaxes the muscles surrounding the vessels and thereby controls blood pressure, among many other functions (*Science*, December 18, 1992, p. 1862). But other research had suggested that it can only do so if it avoids hemoglobin, whose iron-containing heme groups destroy it. "Everybody has looked at hemoglobin as a scavenger of NO," agrees Robert Furchgott at the Health Sciences Center of the State University of New York, Brooklyn. The new results, reported by Li Jia, Stamler, and Celia Bonaventura and Joseph Bonaventura of Duke's Nicholas School of the Environment and the medical center, "are entirely contrary to the dogma in the field," says Stamler.

The finding that hemoglobin can carry a form of NO, releasing it at crucial moments, is "provocative, interesting, and exciting," says Robert Winslow of the University of California, San Diego (UCSD), Medical School. He adds that the finding could be a boon for efforts to turn cell-free hemoglobin solutions into workable blood substitutes; in past tests, patients given such solutions often suffered dangerous rises in blood pressure. "It may be that this will open a new line of research to solve this problem," says Winslow.

The experiments build on Stamler's earlier research on compounds called S-nitrosothiols, or SNOs, which act as souped-up versions of the NO molecule, enhancing its physiological effects. SNOs form in the body when oxidized NO reacts with the highly reactive thiol group—a sulfur and a hydrogen—on the amino acid cysteine. Cysteines are common on proteins, including hemoglobin, which has a pair of them. Their function on hemoglobin "has been a mystery," says Joseph Bonaventura. But the Duke team suspected that SNOs might form on hemoglobin, perhaps when other, smaller SNOs in the bloodstream in effect hand off their NO to the pair of cysteines.

To test the idea, the team incubated hemoglobin in a bath of SNOs. Standard biochemical lore predicted that NO in any form would rapidly react with oxygenated heme groups, inactivating the NO and leaving a positive charge on the hemoglobin. But for SNO, the Duke group found, the outcome was different. SNOs rapidly formed on the hemoglobin's two cysteines, forming SNO-hemoglobin. The free SNOs "were transferring NO groups to cysteine without touching the hemes," says Stamler.

To see whether this effect actually occurs in the blood vessels and can influence their behavior, the team placed chemical "caps" on parts of cell-free hemoglobin molecules, selectively blocking either the hemes or the cysteine thiols or leaving both sites free. They then exposed sections of rat artery to the modified hemoglobin and monitored the artery's response. The blood vessel constricted in response to all three hemoglobin preparations. But the constriction was greatest when neither the cysteines nor the heme were capped. That implied, says Stamler, that the bare hemoglobin was consuming NO, by both binding it at the cysteines and destroying it at the heme.

"The next question was, " says Stamler, "If it occurred in isolated blood vessels, can it occur in the body?" The answer seemed to be yes. The team monitored the levels of SNO bound to hemoglobin in red blood cells collected at various points in rats'

circulation. They found high levels of SNO-hemoglobin in red blood cells on the left side of rats' hearts—where blood has just been oxygenated in the lungs—and low values on the right side; where blood returns from the body. That suggested, says Stamler, that hemoglobin was taking on SNO in the lungs as it was oxygenated and delivering the SNO to the tissue in concert with oxygen release. To confirm that this delivery really does affect blood pressure, the team then injected rats with SNO-hemoglobin. While native hemoglobin, lacking SNO, causes rats' blood pressure to rise, it remained rock steady when the SNO-hemoglobin when in.

Based on biochemical measurements and molecular-dynamics computations, the Duke researchers argue that this regulatory effect reflects the same kind of shape, or allosteric changes that choreograph hemoglobin's familiar functions. Hemoglobin gives up its oxygen and sops up carbon dioxide when, upon reaching oxygen-poor tissues, it undergoes an allosteric change that reduces its affinity for oxygen. The same conformational change also releases SNO, says Stamler. The reverse effect takes place in the lungs, ensuring that hemoglobin takes on SNO at the same time as it binds oxygen and blows off carbon dioxide.

The Duke researchers also found another trigger for SNO release. Each time hemoglobin scavenges an NO in the bloodstream in the standard fashion, the positive charge left on the heme signals to the cysteine residue to release its SNO. "The moment [hemoglobin] gets hit by NO, it gives up one," maintaining a balance and keeping the vessel open, Stamler speculates.

In addition to this surprising insight into hemoglobin's regulatory roles, Vijay Sharm of UCSD notes that the work "could have tremendous significance for blood substitutes. " Simply infusing SNO-hemoglobin and not the SNO-less version, he speculates, might eliminate the blood pressure surge these preparations are notorious for causing. Still, says Sharma, "the devil is in the details," and he, like other researchers, is eager to see all the incriminating details before acting the new picture of hemoglobin's double life. ❏

Questions

1. What is hemoglobin's traditional role in the body?

2. Where is nitric oxide produced?

3. How do hemoglobin and nitric oxide work together in the body?

Answers are at the back of the book.

31

Scientists are finding novel uses for growth hormone (GH) which could have therapeutic value in the area of cardiovascular medicine. Known mostly for its role in somatic development, GH is now being considered necessary for maintaining cardiac mass and performance well into adult life. But GH is not the only growth regulator which affects cardiac growth and function. Insulin-like growth factor I (IGF-I) is also involved as growth hormone's tissue effector. Together, they are opening up new therapeutic designs for improving heart performance in cardiac diseases that commonly lead to heart failure and death.

Cardiac Performance: Growth Hormone Enters the Race

Luigi Saccà and Serafino Fazio

Nature Medicine, January 1996

It is possible that growth hormone can offer therapeutic benefits for treating some forms of heart failure.

Renowned for its pivotal role in somatic development, growth hormone (GH) is involved in a variety of physiological processes, such as the response to stress. Evidence is emerging that points to a novel role for GH—and its local effector, insulin-like growth factor 1 (IGF-1)—in the control of heart morphology and function. What is exciting now is the idea that GH is essential not only for cardiac development but also for maintaining cardiac mass and performance in adult life.[1] Moreover, within the space of a few months, two research groups have reported that GH and IGF-1, respectively, exert a beneficial effect in experimental heart failure.[2,3] These observations have renewed interest in the molecular analysis of heart physiology and have opened whole new therapeutic horizons in cardiovascular medicine.

GH exerts its effects on tissues via both direct and indirect mechanisms; the latter are mediated by locally synthesized IGF-1. In particular, GH induces IGF-1 mRNA expression and increases IGF-1 content in cardiac tissue.[4] Myocardial mass is increased in human models of GH excess.[1] This effect represents cardiac hypertrophy sensu strictu, because myocardial growth is out of proportion to body weight. GH may also enhance myocardial contractility, independent of changes in cardiac morphology. Indeed, contractility is increased in cardiac tissue preparations from animal models of chronic GH excess.[5]

The effects of GH deficiency on the heart are diametrically opposed to those caused by GH excess. In subjects with congenital deficiency of GH, cardiac growth is hampered and myocardial mass is reduced because of ventricular wall thinning. This causes wall stress to rise and cardiac performance to deteriorate.[1] The administration of recombinant human GH to GH-deficient subjects increases ven-

Reprinted with permission from *Nature Medicine*, Vol. 2, No. 1, January 1996.

tricular mass and reduces cardiac wall stress; consequently, both cardiac performance and exercise capacity and normalized.[1]

GH may play a physiologic role in conditions of prolonged stress in which the increased metabolic demands require cardiovascular adaptation. GH also increases skeletal muscle mass and strength.[6] Because of these combined effects on cardiac and skeletal muscle, GH improves physical performance. This finding has encouraged the search for new therapeutic designs in which GH might be used in conditions characterized by decline of skeletal muscle and myocardial performance, such as advanced age and catabolic, chronic illness. On the other hand, the favorable effects of GH on cardiac and skeletal muscle function may be an irresistible temptation for athletes to push their performances beyond measure. A laboratory approach to revealing GH doping is urgently needed.

The idea of GH as a potential activator of cardiac performance would certainly raise more enthusiasm if one could demonstrate that the hormone does in some way intervene in the sequence of events that mediate the cardiac response to work overload. This is a question of the specific role played by IGF-I, the tissue GH effector, in the scenario of the intricate mechanisms controlling cardiac growth. Molecular studies of IGF-I's role in cardiomyocyte growth are nicely intersecting with clinical research on GH control of cardiac physiology. Two observations relevant. First, IGF-I increases the size of cultured cardiomyocytes and induces the transcripts of specific genes, such as myosin light chain-2, skeletal muscle α-actin, and troponin I.[7] Second, IGF-I mRNA expression is increased in the myocardium following pressure overload.[8] Interestingly, IGF-I mRNA increases at an early stage of ventricular hypertrophy and to a greater extent in those myocardial segments where the stress is higher.[9] These data provide a strong argument in favor of a role for IGF-I in cardiac hypertrophy.

Further evidence that IGF-I is involved in cardiac growth comes from observations in IGF-I gene-deficient animals.[10] Myogenesis is compromised in this model system, suggesting a crucial role for IGF-I in developmental physiology. Attempts are under

way to obtain more specific information from knockout animals with selective myocardial deficiency of IGF-I receptor or downstream signaling molecules. In essence, the human model of congenital GH deficiency shows that the role of IGF-I is not marginal or replaceable by redundant mechanisms. These subjects have barely measurable GH and IGF-I levels and carry significant defects in cardiac morphology and performance. The consequences of these abnormalities are not restricted to quality of life, but entail an elevated risk of cardiovascular mortality, mainly due to heart failure.[11]

The possibility of manipulating cardiac growth with GH or IGF-I is opening entirely new strategies in cardiac diseases that lead to heart failure. This is a very common disorder with an epidemiological dimension of 400,000 new cases each year in the United States. An example is provided by dilated cardiomyopathy, the most frequent indication for cardiac transplantation. This disease is marked by ventricular dilation, unaccompanied by adequate wall thickening, thus leading to increased wall stress and severely impaired cardiac performance. These features provide the rationale for expecting a beneficial effect of GH. Indeed, in single case reports a dramatic improvement of cardiac failure was observed after short-term GH treatment in patients with dilated cardiomyopathy secondary to hypopituitarism.[12, 13]

IGF-I was tested as to its ability to affect postinfarction myocardial hypertrophy and dysfunction. In this model, IGF-I induced additional myocyte growth (without affecting capillary density or collagen growth) and improved cardiac dysfunction markedly.[2] More recently, in a similar model of myocardial infarction and heart failure, GH treatment augmented the depressed contractility and increased cardiac output.[3]

The demonstration that therapy with a growth factor is beneficial in experimental heart failure is exciting, both for the potential clinical implications and because it is now clear that the cardiac impact of GH and IGF-I is physiological. This is not the case for other growth regulators they may have divergent effects on cardiac growth and function. For example, angiotensin II is implicated in the growth response to pressure overload but is also regarded as a potential cause of cardiac dysfunction, probably because of excessive collagen synthesis leading to pathologic myocardial remodeling.[14]

The mechanism(s) responsible for these diverse effects of growth factors on cardiac performance represents a major challenge for cardiovascular biology. A characteristic fetal gene program reemerges in response to specific growth factors or hypertrophic stimuli, leading to quantitative and qualitative changes in myocardial proteins. In functional terms, the response may be advantageous, indifferent, or even adverse. In the case of GH, the response is very advantageous because it entails an increase in the force of cardiac contraction and, paradoxically, a lowering of the energetic cost of this process.[5] At molecular level, GH downregulates α-myosin heavy-chain (α-MHC) and reinduces the low ATPase activity β-isoform (β-MHC).[5] Is this the molecular explanation for GH's favorable effect on contractility? Probably not. In other forms of cardiac hypertrophy, such as that secondary to pressure overload, β-MHC expression is equally increased and yet the form of contraction is reduced, in agreement with the lower shortening velocity of β-MHC and its low ATPase activity.[15] Thyroid hormones induce a shift towards α-MHC and improve contractility.[16] Thus, identical changes in MHC isoform expression may be associated with either enhanced or depressed myocardial contractility.

Seeking the functional correlates of individual sarcomeric proteins is too simplistic an approach. Ultimately, what determines the performance of the hypertrophic myocardium is the product of intricate interactions among many reexpressed peptides, which may be contractile, structural and regulatory. Elucidation of these interactions is the necessary prerequisite in the search for novel models in which addition or inhibition of specific growth factors may repress pathologic phenotypes of hypertrophy and orient myocardial remodeling toward physiological advantage.

References

1. Saccà, L., Cittadini, A. & Fazio, S. Growth hormone and the heart. *Endocr. Rev.* **15**, 555-573 (1994).

2. Duerr, R.L. *et al.* Insulin-like growth factor-I enhances ventricular hypertrophy and function during the onset of experimental cardiac failure. *J. Clin. Invest.* **95**, 619-627 (1995).

3. Yang, R., Bunting, S., Gillett, N., Clark, R. & Jin, H. Growth hormone improves cardiac performance in experimental heart failure. *Circulation* **92**, 262-267 (1995).

4. Flyvbjerg, A., Jorgensen, K.D., Marshall, S.M. & Orshov, H. Inhibitory effect of octreotide on growth hormone-induced IGF-I generation and organ growth in hypophysectomized rats. *Am. J. Physiol.* **260**, E568-E574 (1991).

5. Timsit, J. *et al.* Effects of chronic growth hormone hypersecretion on intrinsic contractility, energetics, isomyosin pattern, and myosin adenosine triphosphatase activity of rat left ventricle. J. Clin. Invest. **86**, 507-515 (1990).

6. Jorgensen, J.O.L. *et al.* Beneficial effects of growth hormone treatment in GH-deficient adults. *Lancet* **1**, 1121-1225 (1989).

7. Ito, H. *et al.* Insulin-like growth factor-I induces hypertrophy with enhanced expression of muscle specific genes in cultured rat cardiomyocytes. *Circulation* **87**, 1715-1721 (1993).

8. Wåhlander, H., Isgaard, J., Jennische, E. & Friberg P. Left ventricular insulin-like growth factor-I increases in early renal hypertension. *Hypertension* **19**, 25-32 (1992).

9. Donohue, T.J. *et al.* Induction of myocardial insulin-like growth factor-I gene expression in left ventricular hypertrophy. *Circulation* **89**, 799-809 (1994).

10. Baker, J., Liu, J.-P., Robertson, E.J. & Efstratiadis, A. Role of insulin-like growth factors in embryonic and postnatal growth. *Cell* **75**, 73-82 (1993).

11. Rosén, T. & Bengtsson, B.-Å. Premature mortality due to cardiovascular disease in hypopituitarism. *Lancet* **336**, 285-288 (1990).

12. Cuneo, R.C., Wilmshurst, P., Lowy, C., McGauley, G. & Sonksen, P.H. Cardiac failure responding to growth hormone. *Lancet* **1**, 838-839 (1989).

13. Frustaci, A., Perrone, G.A., Gentiloni, N. & Russo, M.A. Reversible dilated cardiomyopathy due to growth hormone deficiency. *Am. J. Clin. Pathol.* **97**, 503-511 (1992).

14. Pfeffer, M.A. *et al.* Effect of captopril on mortality and morbidity in patients with left ventricular dysfunction after myocardial infarction. Results on the survival and ventricular enlargement trial. *N. Engl. J. Med.* **327**, 669-677 (1992).

15. Lompre, A.M. et al. Myosin isoenzyme redistribution in chronic heart overload. *Nature* **282**, 105-107 (1979).

16. Izumo, S., Nadal-Ginard, B. & Mahdavi, V. All members of the MHC multigene family respond to thyroid hormone in a highly tissue-specific manner. *Science* **231**, 597-600 (1986).

❏

Questions

1. What is considered growth hormone's (GH) primary role in the body?

2. What other physiological roles can growth hormone (GH) and insulin-like growth factor-I (IGH-I) play?

3. What are some of the ways that cardiac growth regulators could be used?

Answers are at the back of the book.

32

Hypertension is a serious national public-health problem, but its causes are largely unknown. When blood pressure becomes dangerously high, it can cause a myriad of problems due to the stress on both the heart and blood vessels. But until complications occur, hypertension is symptomless. Fortunately, once hypertension is detected, proper treatment can reduce the course and severity of the problem. Treating the high blood pressure usually includes reduction in salt intake and the administration of diuretics to lower both salt and water load in the body. Mild to moderate hypertension can be treated with exercise and following the proper diet. Ultimately, long-term control of blood pressure can be controlled with a host of different therapeutic interventions all likely to have comparable benefits.

High Blood Pressure
Harvard Women's Health Watch, July 1996

From time to time, we're conscious of our bodies' inner workings. We may detect a heart beat, a digestive rumble, or a uterine contraction, but we aren't aware of our blood pressure. It is mute, and even when it becomes dangerously high, raising the risk of stroke, heart attack, kidney damage, and vision loss, there may be no symptoms. It is known as a silent killer, but with proper treatment, one that is easily arrested.

Blood pressure is the force exerted against the walls of the arteries and arterioles, vessels that carry blood from the heart. When these muscular walls constrict, reducing the diameter of the vessel, blood pressure rises; when they relax, increasing the vessel diameter, it falls. Blood pressure usually rises in response to exercise and stress; it falls when we're at rest. It also mounts when blood volume—the amount of fluid in the vessel—increases.

A blood pressure measurement is a "snapshot" of one's blood pressure at a given moment. The person who takes your pressure will wrap a cuff around our upper arm and inflate it until it compresses an artery, stopping blood flow. He or she will then slowly release the air from the cuff while listening to the artery through a stethoscope. The clinician will hear the blood flow into the vessel when the pressure in the cuff equals that in the artery. The reading at that point represents the systolic pressure, or the pressure when the heart muscle is contracting. The reading taken when sounds are no longer heard represents the diastolic pressure or the pressure when the heart muscle is relaxed. The systolic pressure is the top number, and the diastolic pressure is the bottom number of the reading.

A normal blood pressure reading is an indication that the arteries and heart are dilating and contracting properly to deliver blood to all parts of the body. Your clinician might want to take several readings—while you are standing, sitting, and lying down—to see how you blood pressure responds to change. If your doctor suspects that the office reading is higher than it might normally be due to stress of the office visit—a phenomenon called "white coat hypertension"—he or she may recommend a 24-hour reading taken with a home monitor.

A high blood pressure reading is usually a sign that the vessels cannot relax fully. They remain somewhat constricted, requiring the heart to work harder to pump blood through them. Over time, the extra effort can cause the heart muscle to become enlarged and eventually weakened. The force of blood pumped at high pressure can also produce small tears in the lining of the arteries, weakening the arterial vessels. The evidence of this is most pronounced in the vessels of the brain, those to the kidneys, and the small vessels of the eye.

Treating High Blood Pressure

Hypertension is defined by an elevation of diastolic pressure, systolic pressure, or both. A diastolic pressure blow 85 is considered normal; 85-89, high normal; 90-104, mild hypertension; 105-114, moderate hypertension, and greater than 115, severe hypertension. Systolic pressures of 140-159 are defined as borderline systolic hypertension and those over 160 as systolic hypertension. A diastolic pressure over 90 and/or a systolic pressure over 160 has been consistently linked with an increased risk of stroke, heart failure, coronary artery disease, kidney damage, and vision loss.

The risk of hypertension increases with age as arteries lose elasticity and become less able to relax. About 70% of women between ages of 65 and 75 have high blood pressure, particularly high systolic pressure.

African-American women are at even higher risk than Caucasians. Those who develop high blood pressure do so earlier in life and have more severe elevations. They also experience a greater increase in the risk of stroke or heart disease as a result. By age 65 almost 80% of black women have high blood pressure.

As a rule, the treatment depends on how high blood pressure is and whether symptoms of heart disease, kidney damage, or vision loss are present. The following are recommendations for each level of hypertension:

• *High normal.* Try to reduce the known risk factors for high blood pressure. Diet and exercise are key.

There is mounting evidence that the minerals calcium, potassium, and magnesium are protective, especially when obtained through a diet rich in lowfat dairy products, green leafy vegetables, fruits and grains. Restrict sodium to 2,400 mg a day. (Although recent studies indicate that only a fraction of people with elevated blood pressure are sodium sensitive, the only way to determine if you are is to minimize salt consumption and see if your blood pressure falls.) Increase dietary fiber to 25 grams daily and limit alcohol consumption to one drink a day.

Exercise can also be beneficial. Try to get at least 30 minutes of moderate exercise every day. Other techniques, such a relaxation and biofeedback, though appealing in theory, haven't reduced blood pressure significantly in clinical trials.

Following the proper diet and exercise routine should also help to keep weight down. Women who are more than 10% above ideal weight can often lower their blood pressure by losing the extra pounds.

If you're past menopause, hormone replacement therapy shouldn't increase your blood pressure. Although the results are inconclusive, estrogen, which is thought to relax blood vessels, has been associated with reductions n blood pressure in some studies.

• *Mild hypertension and borderline systolic hypertension.* Because drug treatment may lead to a lifetime on medication, which in itself may have some complications, most experts believe that the dietary and life-style changes recommended for people with high normal blood pressure should be the first approach. If these fail to lower blood pressure, drug treatment is recommended.

• *Moderate and severe hypertension.* This includes isolated systolic hypertension (systolic hypertension with normal diastolic pressure), which is common in older women. Drug treatment is recommended.

Drug Therapy

The following classes of drugs are commonly used for high blood pressure:

• *Diuretics,* also known as "water pills," which lower pressure by eliminating sodium and water and thus reduce the volume of blood in circulation, are often the first approach. They may be beneficial for women who have benign bloating or swelling and are particularly effective in African-American women. Thiazide diuretics, which increase calcium reabsorption by the kidneys, can also help to prevent osteoporosis. Some diuretics remove potassium from the body, so those who use them may need to eat more potassium-rich foods or take a supplement.

• *Beta blockers* relax not only the blood vessels, but the heart muscle, producing a decline in the force of contraction and reduction in heart rate. These drugs may be a good choice for women who are anxious but aren't recommended for those who like to exercise vigorously or for people with asthma. They may not be as effective for black women as for whites.

• *Angiotensin-converting enzyme (ACE) inhibitors* are highly selective drugs that interrupt a chain of molecular messengers that constrict blood vessels. They can improve cardiac function in people with heart failure and are good choices for those with diabetes or early kidney damage. Rashes and coughs are common side effects.

• *Calcium channel blockers* inhibit calcium from entering the blood vessel walls where it powers the process that constricts blood vessels. Because recent studies have indicated that short acting calcium blockers increase heart-attack risk in some people, they should be used cautiously.

In the past, drug therapy began with diuretics. If those drugs didn't bring pressure down, others were added to the regimen. Today, there are so many different agents available that a single drug treatment is likely to be adequate for most women. When the original drug isn't effective, another is usually substituted.

Diuretics and beta-blockers are the only drugs that have been extensively studied and shown to reduce the risk of heart disease, stroke, and death in clinical trials. however, many others effectively reduce blood pressure and are likely to have comparable benefits.

Secondary Hypertension

In about 5% of cases, high blood pressure fails to respond to medication. In these instances, it may be due to a specific medical problem, such as an adrenal tumor or kidney disease. It will usually resolve if the underlying condition is successfully treated.

Oral contraceptive use may cause blood pressure to rise in some women, although today's birth control pills, which obtain lower doses of estrogen and/or progestin, are less likely to have this effect than earlier formulations. High blood pressure due to the pill usually returns to normal when women stop taking it.

Preeclampsia, also called pregnancy-induced hypertension, affects many systems in the body. Its cause is unknown. Frequent prenatal checkups to monitor blood pressure and protein in the urine—signs of preeclampsia—are important. Although hospitalization or bed rest is often the treatment of choice, antihypertensives are occasionally prescribed. Blood pressure usually returns to normal following delivery. ❑

Questions

1. What is blood pressure?

2. Why does the risk of hypertension increase with age?

3. How do diuretics work in the body to lower blood pressure?

Answers are at the back of the book.

---

33 *Small, and quite often forgotten, the thyroid performs a very important function for the body regulating our basal metabolism. Through the diet, we consume iodine which is readily taken up by the thyroid gland. After a series of conversions, it enters the bloodstream bound to protein in the form of the hormones triiodothyronine (T3) and thyroxine (T4). Once T3 and T4 enter the cells of the body, they control the production of certain gene products that affect our energy requirements. Both the hypothalamus and pituitary help to control the thyroid's release of these hormones. Although disease related to the thyroid and its output can create chronic conditions for the patient, it can be successfully treated.*

Thyroid Diseases

Celeste Robb-Nicholson, M.D.

Harvard Women's Health Watch, January 1995

The thyroid occupies a low position in the collective consciousness, yet this small gland influences virtually every cell in the body. The hormones it secretes into the bloodstream play a vital role in regulating our basal metabolism—the rate at which we convert food and oxygen to energy.

How the Thyroid Works

In some respects the thyroid is like a small factory. It is the site where iodide—the byproduct of the iodine we absorb in food—enters from the bloodstream, is converted back into its original form, and then is welded to proteins to form the hormones triiodothyronine (T3) and thyroxine (T4). T3 enters the cells directly and activates the genes that direct the production of certain proteins. T4 circulates in the blood until T3 is needed; then it is taken into the cells and converted to T3.

The thyroid's output is determined by the hypothalamus, a regulatory region of the brain. The hypothalamus sends thyrotropin-releasing hormone (TRH) to the pituitary—a peanut-sized gland at the base of the brain—which is entrusted with maintaining thyroid-hormone production at genetically determined levels. When the pituitary senses that the supply of thyroid hormones threatens to drop below those levels, it emits thyroid-stimulating hormones (TSH) to trigger the production and release of T3 and T4.

Because the thyroid hormones influence so many types of cells, it's not surprising that when something goes awry we may experience an array of symptoms that seem unrelated. Changes in weight, bowel habits, heart rate, hair growth, temperature tolerance, and menstrual cycles—when they occur together—can signal a thyroid condition.

Our susceptibility to thyroid disease is largely determined by the interaction of our genetic make up, age, and gender. Women, particularly those with a family history of thyroid disease, are much more likely to have thyroid trouble than men. Fortunately, although most thyroid conditions can't be prevented, they respond well to treatment.

When Hormone Levels Are Too Low

Hypothyroidism, too low a level of the hormones, is the most common thyroid disease in this country, with more than 50% of cases occurring in families where thyroid disease is present. Although as many as one woman in 10 develops hypothyroidism, its symptoms appear so gradually that many are unaware that they have it.

As thyroid hormone level fall, the basal metabolic rate drops, resulting in a decreased heart rate, and increased sensitivity to cold, sluggishness, constipation, a slowdown in hair and nail growth, and a weight gain that is limited to 10-15 pounds. (Contrary to widespread misconception, hypothyroidism is not responsible for obesity.) Because blood hormone levels are consistently low, the pituitary send a steady stream of TSH to the thyroid in an attempt to encourage production. This constant stimulation sometimes causes it to enlarge, creating a goiter.

Hypothyroidism may occur in people who take lithium or certain asthma medications, or who have undergone radiation therapy for certain cancers, such as Hodgkin's disease or throat cancer. Iodine deficiency, once a common cause of hypothyroidism, is no longer a problem in this country; most of us get more than enough iodine from salt, bread, milk, and certain prescriptions and over-the-counter medications.

Hashimoto's thyroiditis is among the most common causes of hypothyroidism in women over 50. It is an inherited condition in which the body produces antibodies against its own thyroid tissue, eventually damaging the gland so much that it can no longer produce enough T3 and T4.

The most accurate test for hypothyroidism measures levels of TSH in the blood. A high pituitary level indicates that the pituitary is signaling the thyroid to produce more hormone in an attempt to compensate for too little T3 and T4.

Regardless of the form of hypothyroidism, the condition can be successfully treated with tablets containing thyroxine, or T4, which can be regulated to maintain adequate levels of circulating hormone. Blood levels of the hormones should be checked periodically and the dosage adjusted when necessary, because needs may vary with age, body weight, and use of certain medications. Most patients need to take the thyroxine for the rest of their lives.

When Levels Are Too High

As might be expected, the symptoms of hyperthyroidism are the opposite of those that characterize hypothyroidism—a modest weight loss despite a normal appetite, bouts of mild diarrhea or frequent bowel movements, heat sensitivity, trembling, increased heart rate, and emotional changes. Women with hyperthyroidism have lighter periods and a higher rate of infertility and miscarriage.

There are several conditions that can cause hyperthyroidism, but about half of all cases are the result of *Grave's disease*, which, like Hashimoto's thyroiditis, is hereditary. It is seven to nine times more common in women than in men. It is also an autoimmune disease, but instead of destroying the thyroid tissue, the antibodies in patients with Graves' disease mimic the effects of TSH, triggering thyroid enlargement.

Like those with other forms of hyperthyroidism, people with Graves' disease often take on a "wide-eyed" look, caused by retraction of the upper lids. Less commonly, the immune reaction responsible for Graves' disease creates swelling in the muscles of the eyes, causing them to bulge, and produces raised, plaque-like areas on the skin of the legs.

In those with a family history of Graves' disease, the condition is most likely to strike between the ages of 20 and 40, sometimes in response to hormonal change, such as pregnancy. It may also occur after severe emotional stress.

Evaluation for Graves' disease includes a blood test for low level of TSH and often an iodine uptake test, in which the patient swallows a solution containing radioactive iodine. The physician then uses a scanning devise to measure the amount of iodine that has been absorbed by the thyroid; an elevated level further confirms that the gland is overactive.

Younger patients with mild Graves' disease can be successfully treated with drugs, such as propylthiouracil (PTU) or methimazole, for a period of 6 to 12 months. These drugs make it impossible for the thyroid to use iodine and thus block hormone production. Because antithyroid drugs have several undesirable side effects, many doctors now treat patients initially with radioactive iodine, which destroys thyroid tissue. The dose used to treat hyperthyroidism is much larger than that used to diagnose it, but it passes out the body through the urine in 48 hours.

Regardless of the method of treatment, high levels of circulating thyroid hormone may remain in the blood for several weeks, so another drug called a beta-blocker is often used to counteract the thyroid hormone's effects on the heart and vascular system. A majority of those treated for Graves' disease will eventually have low thyroid function and require thyroid pills.

Other Forms of Hyperthyroidism

A single lump, or nodule, which often develops in the thyroids of people who received low-dose radiation in childhood, can sometimes secrete high levels of hormone. When the nodule's production reaches an excessive amount, symptoms of hyperthyroidism begin.

When there is more than one nodule, the resulting condition is called *toxic multimodular goiter*. It usually occurs in people over age 65, and is responsible for about 25% of the cases of hyperactive thyroid.

Although beta-blockers and antithyroid drugs can alleviate the symptoms caused by thyroid modules, the only permanent cure is radioactive iodine to destroy the active thyroid tissue. It also eventually necessitates thyroid supplement.

Radioactive iodine therapy has been so successful that thyroidectomy, or removal of the gland, is now rarely used to treat hyperthyroidism. Surgery is usually reserved for patients with goiters that interfere with breathing or swallowing or are unsightly.

Subacate thyroiditis is a less common form of hyperthyroidism caused by a viral infection. In addition to flu-like symptoms, the patient experiences an inflammation of the thyroid, which allows excess hormone to leak out into the bloodstream. Although the effects of this form of hyperthyroidism can also be countered with beta-blockers, the symptoms and the underlying infection eventually clear up with treatment. A painless form of thyroiditis occasionally occurs in women who have recently given birth.

Thyroid Cancer

Rarely, a lump in the thyroid will have more serious implications. Hoarseness, difficulty swallowing and rapid enlargement of a nodule that is not accompanied by symptoms of hypo- or hyperthyroidism may be signs of thyroid cancer. Fortunately, this form of cancer is rare, occurring in 25 people per million, and usually curable. It is most common in those who have received radiation to the head or neck in childhood or who have a family history of thyroid cancer.

When doctors evaluate a thyroid lump, they perform blood tests for hormone levels and autoantibodies to rule out other conditions. A thyroid scan can determine if the nodules fail to take up radioactive iodine and are therefore "cold" or nonfunctional, which raises the suspicion of cancer. Ultrasound or needle biopsy can indicate whether a cold nodule is a benign fluid-filled cyst or a potentially malignant one. Treatment for thyroid cancer entails removal of all or part of the thyroid, and if cancer has metastasized, therapy with radioactive iodine to destroy the remaining malignant cells is recommended. ❏

Questions

1. What is the function of the thyroid?

2. What is TSH?

3. What is the most common form of thyroid disease in this country?

Answers are at the back of the book.

---

34 *Nothing sounds sweeter to couples trying to bear children than the two words, "You're pregnant." But for many couples with infertility the prospects seem dim as month after month passes by with no baby. For this very reason, couples turn to assisted reproduction techniques to help them conceive. Before entering a program, your gynecologist will gather as much information about you and your reproductive health a possible, as well as from your partner. When standard therapy fails, couples can elect for more high-tech approaches to infertility such as in vitro fertilization, gamete intrafallopian transfer and zygote intrafallopian transfer which all involve some method of embryo transfer. Assisted reproduction can be exhausting and expensive but for many it's a price worth paying to conceive a baby.*

Assisted Reproduction

Celeste Robb-Nicholson, M.D.

Harvard Women's Health Watch, April 1995

Most of us grew up with the notion that it's extremely easy to get pregnant. Needless to say, learning that we may not by able to bear children can come not only as a great surprise but a major blow.

A few decades ago, the only thing one could do for infertility was to suffer in silence. Today, not only are a host of physiological and social support services available, but several effective treatments as well. Although such high-tech approaches to infertility as *in vitro* fertilization (IVF), gamete intrafallopian transfer (GIFT), and zygote intrafallopian transfer (ZIFT) may be exhausting and expensive, increasing numbers of couples who are able to keep trying will eventually conceive.

When to Consider Getting Help

Your chance of successful infertility treatment declines rapidly after the mid-30s. If you're in that age range and have been trying to get pregnant for 6 months to a year, you should probably talk to your doctor. (Women under 35 should try for a least a year before seeking help.) Most fertility programs have maternal age limits of 42 or 43, and very few accept women over 45 because their chances of becoming pregnant are so poor.

Your doctor will want to take your medical history, do a physical examination, and draw blood to be tested for hormone levels. Of course, you're not alone in this enterprise, so your male partner will need an evaluation as well.

If your periods are irregular, your blood may be tested at the beginning of your menstrual cycle for follicle-stimulating hormone (FSH), which causes the egg-containing follicle to ripen. High FSH levels indicate than the pituitary gland is producing large amounts of FSH in an effort to get the ovary to release an egg. A blood test for progesterone, which is produced by the empty ovarian follicle (corpus luteum), can be performed later in your cycle to determine if an egg has developed and possibly been released. Another indication that ovulation may be occurring is a mid-monthly decline in body temperature followed by sustained rise, so you may be asked to record your temperature every morning upon awakening and to plot it on a chart. Not all women experience the temperature drop at ovulation, so you may also be asked to use a home urine test for levels of luteinizing hormone (LH), which rise just before the egg is released, usually between days 10 and 18 of the menstrual cycle.

If these tests indicate that you're not ovulating regularly, your gynecologist may prescribe one of several drugs to stimulate the ovaries, such as clomiphene citrate (Clomid) in pill form, or either human menopausal gonadotropin (Pergonal) or FSH (Metrodin), given as daily injections in the buttocks. All can have unpleasant side effects.

Your male partner will need to donate a semen sample for analysis. If the results are abnormal, he'll be required to repeat the donation, since sample quality can vary widely.

If both you and your partner's fertility evaluations are within normal boundaries, and you have been having intercourse regularly during your ovulatory periods, your doctor will probably suggest a postcoital test. This will require having vaginal specimen taken within a few hours of intercourse. The sample will be examined under a microscope to determine the number and condition of the sperm that remain and will give your doctor an indication of whether your vaginal mucus is inhospitable to the sperm or you are producing antibodies that disable them.

If your doctor suspects that you are ovulating regularly, but your fallopian tubes are blocked, he or she will probably advise a *hysterosalpingogram* (HSG). In this procedure, dye is infused through the cervix into the uterus and fallopian tubes so that they can be viewed on an x-ray. If the results of the above tests are normal, or if your doctor suspects endometriosis or pelvic adhesions, your physician will probably perform a *laparoscopy* in which a tiny camera is inserted into the abdominal cavity under general anesthesia. This technique allows the doctor to see inside the abdominal cavity and, if necessary, to remove any lesions. (See *HWHW*, September 1994.)

For 10-15% of couples, no reason for infertility can be found. If less invasive procedures like ovarian stimulation and intrauterine (artificial) insemination aren't effective, these couples, like those for whom the reasons for infertility are known, may benefit from IVF, GIFT, or ZIFT.

Assisted Reproduction Techniques

The first step in any of the IVF procedures is to "hyperstimulate" the growth of ovarian follicles with Clomid, Pergonal, or Metrodin. You may also be given a gonadotropin releasing hormone (GnRH) agonest, such as Lupron, for several days at the beginning or end of the cycle to suppress your natural hormonal surges. Hyperstimulation causes several eggs to ripen, increasing the number of opportunities for fertilization.

The development of your follicles will be monitored carefully with frequent blood tests and ultrasound. When they ripen, you'll receive an injection of human chorionic gonadotropin (hCG), to trigger the release of eggs.

Retrieving the Eggs

About 34-36 hours after receive the hCG injection, you'll need to return to the clinic so that the eggs can be retrieved. If you're not using donor sperm, your male partner must go along to provide a semen specimen.

The eggs are usually retrieved by a procedure in which the surgeon inserts an ultrasound probe into your vagina. After administering an anesthetic, he or she will pass a hollow needle through the vaginal wall, puncture the ripened ovarian follicles, aspirate the egg-containing fluid, and withdraw the needle. Alternatively, a laparoscopic procedure, which requires general anesthesia, is used for GIFT or ZIFT for further evaluation of the pelvis.

What happens next depends on the method of embryo transfer you're using. All have similar pregnancy rates.

• *IVF* is usually used for women whose infertility results from obstructed fallopian tubes, endometriosis, or cervical mucus that is inhospitable to sperm. It is also used for some couples with unexplained infertility. The harvested eggs are examined by an embryologist, who places them in a culture dish along with the sperm. When it's evident fertilization has occurred, three or four embryos are transferred via the vagina into the uterus. Using several embryos increases the chance of pregnancy, but it also increased the likelihood of multiple births. Any additional embryos can be frozen for use in a subsequent attempt if the couple wishes.

• *GIFT.* The eggs are retrieved laparoscopically, and then, during the same operation, mixed with the sperm and placed in one or both fallopian tubes. GIFT requires at least one functioning tube and is most often used in cases of unexplained infertility, cervical damage, or mild endometriosis. It is also appropriate for couples in whom the sperm cannot penetrate the cervical mucus.

Unlike IVF, GIFT requires general anesthesia and doesn't allow fertilization to be documented before the egg and sperm are placed in the tube.

• *ZIFT,* an infrequently used assisted reproduction method, is similar to GIFT, except the fertilization occurs in a culture dish before the two-celled embryos, or zygotes, are returned to the fallopian tubes. It is usually reserved for women with severe cervical damage who have at least one working fallopian tube. Both ZIFT and GIFT carry an increased risk of tubal pregnancy.

• *ICSI,* or intracytoplasmic sperm injection, is used for couples in whom sperm count is very low or when motility is extremely poor. In this procedure, a single sperm is injected into an egg, which is then incubated to allow fertilization to take place. The procedure is still experimental, and few clinics are experienced in performing it.

Embarking on assisted reproduction is a major undertaking for any couple. It's wise to determine the cost of treatment and associated expenses, such as travel, in advance. Because insurance coverage is highly variable, check with your insurer to determine how much of the bill you'll have to pay.

While you are undergoing evaluation and treatment, it is likely to become the central focus of your life. Like most couples with infertility, you may benefit from seeing a counselor or joining a support group to help you identify options you may not have considered, such as adoption or deciding not to have a child at all. ❑

Questions

1. What is FSH and what does it do?

2. In what percentage of couples can no reason for infertility be found?

3. What are the success rates for the various embryo transfer methods?

Answers are at the back of the book.

35

Maleness has always been synonymous with the hormone testosterone. Rightly so, far when the gonads receive a signal to develop testes, they begin to produce the male hormone. But testosterone is not enough. According to researchers, a second hormone must act in concert with testosterone to produce a normal, fertile male. To make a male, the hormone müllerian-inhibiting substance (MIS) is produced in the gonads and released to actively suppress the development of female reproductive tracts. Both male and female possess the MIS gene for producing the hormone. When researchers mutated the MIS gene, it produced some very interesting results. In the end, to achieve the proper sex development requires a balance between these hormones.

Not by Testosterone Alone

Sarah Richardson

Discover, April 1995

There is another hormone that is crucial to manhood—and also to male mousehood. Without it, a male mouse has plumbing problems.

Making a man is the complex task of culture. Making a male—nature's job—is also none too simple. The first requirement is the sex-determining gene on the Y chromosome, called the SRY gene: it enables a fetus to build testes, which manufacture the sine qua non of maleness, testosterone. But testosterone alone is not enough; a second hormone is required to produced a normal, fertile male. Molecular geneticist Richard Behringer and his colleagues at the University of Texas M.D. Anderson Cancer Center in Houston have shown just how important this hormone is in male mice. When it is absent, the males develop as infertile pseudohermaphrodites—meaning they have the internal reproductive ducts of a female as well as those of a male. "They look like a regular male," say

Behringer. "You wouldn't know unless you look inside that something is funny."

In mammals, Behringer explains, male and female embryos are identical in the early stages of development. Each fetus has one set of gonads—reproductive glands that develop into either testes or ovaries. IF the gonads receive the signal from the SRY gene, they develop into testes. Other wise they become ovaries.

But this unisex blueprint includes *two* sets of reproductive ducts that extend from the gonads to the embryonic urinary tract. One set of ducts can develop into three female parts: the oviducts (or fallopian tubes in humans), which convey the egg to the uterus; the uterus itself; and the upper portion of the vagina. Very close by lies a second set of ducts that can develop into a male transport system; the epididymis, where sperm cells mature; the vas deferens, the tube that carries mature sperm cells from testes to penis; and the seminal vesicle, which secretes a fluid into the vas deferens that helps rush the sperm on their way "It's not like the gonads, where you have one pair that have to become one or the other," says Behringer. "You've got two sets of ducts, and you've got to make one set differentiate and one regress, depending on whether it's a male or female. That takes a bit more coordination."

Making a female is still relatively straightforward, though: in the absence of testosterone the embryonic male ducts wither away. But researchers have long suspected that to make a male, a second hormone must actively suppress the development of the female reproductive ducts, known as müllerian ducts. The hormone, which is made in the gonads, is called müllerian-inhibiting substance, or MIS. Curiously, the MIS gene is on an ordinary chromosome, not on the Y, so both males and females have it.

To pinpoint the role of the MIS gene in embryonic development, Behringer and his colleagues knocked it out: they mutated the gene in embryonic mouse cells, transferred the embryos into a host mother, and bred the resulting offspring until they had mice in which both copies of the MIS gene were mutated. Then they studied the mutants. The mutant females, the researchers found, didn't miss MIS; the gene is turned off in a female embryo anyway, so the mutation had no effect. The mutant males also appeared to be normally equipped. But they had some extras—oviducts, uterus, and the upper bit of a vagina. The lack of MIS had permitted the female ducts to flourish.

Surprisingly enough, that didn't interfere much with the males' conduct. They could still produce normal sperm and copulate with females. What they couldn't do was fertilize a female. Although the mutant males manufactured normal amounts of sperm, 90 percent couldn't father offspring. The presence of the female reproductive duct, Behringer suspects, somehow blocks the path of the sperm cells before they reach their destination. "The two ducts come together," he says, "and you have a plumbing problem."

The same problem has been observed in humans, although it is extremely rare: a man may have fallopian tubes and a uterus, and be sterile, as a result of a mutation in the MIS gene. "What fascinates me is how this delicate balance in normal development is maintained." say Behringer. "At the molecular or embryological level, there's not much difference between men and women. We all start off similar, and a little hormone here or there and you end up being a male or female." ❑

Questions

1. What gene signals the gonads to develop testes and make testosterone?

2. What function does the mullerian-inhibitory substance have in the male embryo?

3. On which chromosome is the MIS gene found?

Answers are at the back of the book.

134

36

No pain, no gain. Behind the motto of many athletes who train extensively is the fear that without proper conditioning, they could suffer the consequences of electrolyte, renal and volume disturbances from the most trivial level to the most life-threatening. The metabolic stress of exercise can manifest itself in many forms after the loss of water and electrolytes takes place from the body with complication resulting in muscle cramps, heat exhaustion, and heatstroke. Most serious at the level of the kidney is the risk of developing acute renal failure related to exercise in an underconditioned athlete. Along with better conditioning, athletes must rely on adequate hydration to protect themselves from metabolic distress during extreme exercise and ultimately enhance athletic performance all around.

Exercise-Induced Renal and Electrolyte Changes: Minimizing the Risks

Steven Fishbane, M.D.

The Physician and Sportsmedicine, August 1995

A 34-year old woman was thrilled to overcome hot, humid conditions to finish a half-marathon. It was a remarkable achievement, because she had trained for only 2 weeks. As she crossed the finish line she was surprised to find that she was not overly tired. Aside from some calluses on her feet, she experienced little discomfort, although she later recalled some lightheadedness when she walked to the sponsor's tables to get a soda after the race.

That night, however, she noticed that the color of her urine was dark red, and by the afternoon of the next day she had stopped urinating. Several hours later she experienced severe pain in both thighs, and after several episodes of nausea and vomiting, she wen to a local hospital. She was found to be an acute renal failure, with biochemical evidence of rhabdomyolysis. Ins pite of her severe renal

condition, she eventually recovered uneventfully, never required dialysis, and was discharged from the hospital with near normal renal function.

This case illustrated a rare complication of extreme exercise in an athlete who was not well conditioned. Minor homeostatic disruption, such as moderate hypovolemia, hyperkalemia, lactic acidosis, and hematuria, are much more commonly associated with exercise. Sports medicine physician need to stay alert for both mild and severe renal and electrolyte changes.

Hypovolemia and Electrolyte Depletion

Deflects in plasma volume and electrolyte stores occur routinely with physical activity. Exercise causes hypovolemia in proportion to the quantity of water and sodium lost in sweat. When a person exercises

Steven Fishbane, "Exercise-Induced Renal and Electrolyte Changes," *The Physician and Sportsmedicine*, Vol. 23, No. 8, August 1995. Copyright 1995, McGraw-Hill Inc. Reprinted with permission of McGraw-Hill, Inc.

strenuously on a hot day, the volume of sweat lost can exceed 2 L/hr.[1] In addition, plasma osmolality and sodium concentration are increased because sweat is hypotonic relative to plasma. Both the hypovolemia and hyperosmolality are potent stimuli of thirst, which normally leads to correction of both abnormalities when hypotonic fluids are consumed. During exercise, however, fluid consumption often does not keep up with rapid volume losses.

One circumstance in which severe hypovolemia may develop is in the "weekend warrior" who competes in athletics while consuming large quantities of alcohol. The alcohol blocks the effect of antidiuretic hormone in the kidney, preventing maximal concentration of the urine and exacerbating exercise-induced hypovolemia.

The loss of volume during exercise is accompanied by a significant increase in blood flow to muscle and skin.[1] The combination of volume lost as sweat and the redistribution of blood flow to peripheral pools significantly decreases effective circulating volume, stroke volume, and cardiac output.[2] The ability to tolerate this cardiovascular stress depends on the individual's cardiovascular function and level of conditioning. Physical training and heat acclimation both improve the response to hypovolemia.

Hypertension during exercise is prevented by intense vasoconstriction of the splanchnic vascular bed.[3] Alpha-adrenergic receptor blockers (commonly used to treat hypertension) block this protective mechanism and can predispose a patient to hypotensive episodes during exercise. These medications should be used with care in patients who exercise strenuously.

The obligatory loss of volume during exercise is accompanied by losses of electrolytes such as potassium. Potassium depletion can develop after exercise-associated hyperkalemia has reversed with rest. The usual potassium concentration is sweat is 8 mEq/L; therefore, to develop postexercise hypokalemia, total volume losses must be substantial. It has been suggested that potassium depletion may be important clinically, playing a role in the development of heatstroke.[4] Approximately 50% of patients with heatstroke are found to be hypokalemic.[5]

Complications. Loss of water and electrolytes leads to the complications of exercise that are most likely to bring a person to medical attention, including muscle cramps, heat exhaustion, and heatstroke. Exercise-induced *muscle cramps* are common and occur even in well-trained athletes. They can be treated with rest, fluids, and electrolyte replacement. Prevention requires adequate fluid and electrolyte intake before and during exercise.

With *heat exhaustion*, patients experience weakness, headache, nausea, confusion, and faintness. They may have mild hypotension and tachycardia with normal body temperature, but body temperature often elevated. As with muscle cramps, rest, fluids, and electrolyte replacement are essential. Replacement can be accomplished orally, but intravenous fluids are sometimes required.

Heatstroke is the most serious of the heat-induced syndromes. It occurs following exercise most commonly in older people who have underlying diseases and rarely in young, healthy individuals. This syndrome is a medical emergency and requires immediate recognition and treatment. Its symptoms may be similar to those of heat exhaustion; however, body temperature is elevated, and the patient may be near death. Acute renal failure, rhabdomyolysis, and lactic acidosis may all occur with heatstroke. Treatment involved measures to induce immediate cooling, including infusion of chilled intravenous fluid, massage of the extremities to alleviate vasoconstriction, and use of a fan to promote convective loss of heat.

Prevention. The optimal fluid replacement for volume and electrolyte losses occurring with exercise is controversial. Perhaps the most important principle is to replace as much of the fluid and electrolytes lost as possible, preferably during the exercise itself. Replacing volume losses and preventing hyperosmolality can increase cardiac output, enhance athletic performance, and lower rectal temperature.[6]

At least 1,250 mL of 8% carbohydrate-containing fluids can be absorbed from the intestines per hour.[7] Theoretically it should, therefore, be possible to replace a significant proportion of lost volume in all but the most extreme conditions. In practice, however, the amount of fluid consumed is frequently inadequate. This is particularly true with runners, who usually do not consume more than 500 to 1,000 mL per hour. In part, this may be explained by abdominal discomfort brought on by fluid consumption,[8] however, an important component in many is simply a reluctance to drink fluids during exercise.

Replacement fluids should be hypotonic relative to plasma, to match volume losses as sweat. This will help prevent hyperosmolality. Water by itself, however, may rarely be problematic: Cases of postexercise hyponatremia due to excessive water intake have been reported.

Several sports drinks are marketed as an improved method to replace water and electrolyte losses. The primary ingredients in these drinks are usually water and sugar in the form of corn syrup. The sodium content of typical brands is 10 to 20 mEq/L. Potassium tends to be present in very small concentrations (2 to 4 mEq/L), as are other electrolytes such as calcium and magnesium. The beverages also often contain water-soluble vitamins.

The advantage of these liquids over water is that they provide rapid electrolyte and energy replacement (typically, 200 to 300 kcal/L). Carbohydrates in sport drinks may have a benefit because relatively low concentrations (6% to 8%) can serve as both an energy source and a thirst stimulant.[8] There is, however, no clinical proof that these products are necessary. Eating a small snack of almost any type, in addition to drinking water, provides adequate replacement of lost volume and solutes.

Hyperkalemia

Paradoxically, as water and electrolytes (including potassium) are lost from the body during exercise, minor increases in plasma potassium concentration are common. With relaxed walking, increases of approximately 0.3 mEq/L are typical. With more vigorous aerobic exercise, the elevation could reach 0.7 to 1.1 mEq/L.[8-10] With severe exertion (such as marathon running), plasma potassium could increase by as much as 2 mEq/L.[11]

In general, mild hyperkalemia is well tolerated and has no clinical significance. The usual symptoms related to severe hyperkalemia—muscle weakness and cardiac arrhythmias—are only very rarely seen after exercise. Once could speculate, however, whether the increased potassium concentration might play a role in reported episodes of sudden death during exercise.[12] It is possible that in a diseased heart with a predisposition to arrhythmias, hyperkalemia might act as an irritant, thereby decreasing the threshold for malignant ventricular rhythms.

The increased plasma potassium concentration that occurs during exercise results from the release of potassium from muscle cells, followed by delayed cellular reuptake.[13] The slowed reuptake appears to be due to reduced activity of the membrane Na+-K+-ATPase. The increased local tissue contraction of potassium may have a beneficial effect, because it leads to vasodilation, which increases blood flow.[14] This provides improved oxygen and nutrient delivery to starved muscles cells and increased removal of accumulated lactic acid.

The increase in plasma potassium concentration generally resolves within 5 minutes after resting.[15] Potassium released from muscle is rapidly reclaimed by cells, so that by the time the physician sees a patient with an exercise-induced injury or illness, the plasma potassium concentration will usually not be elevated.

One situation in which exercise could induce severe hyperkalemia is in patients taking beta-blockers. Beta$_2$ receptor stimulation normally facilitates the movement of potassium into cells. Blockage of these receptors prevents reuptake of potassium released from muscle cells during exercise. Plasma potassium concentrations can increase by as much as 1.5 to 4.0 mEq/L under these conditions.[18] When a patient develops cardiac arrest temporally related to exercise, the physician should check to see if beta-blockers had been prescribed.

The magnitude of hyperkalemia with exercise is attenuated in well-conditioned athletes.[14] Conditioning enhances Na+-K+-ATPase activity, probably allowing more rapid reuptake of released potassium.

Hyponatremia

Because loss of sweat elevates plasma osmolality and serum sodium concentration, it is surprising that cases of hyponatremia have been reported with exercise. Since sweat is hypotonic, the only way that hyponatremia can develop is if the athlete consumes large amounts of water.[17] Even in this case, the kidneys should still be able to protect against hyponatremia by producing a dilute urine. Something, therefore, interferes with the kidney's powerful diluting capacity in these patients. One possible mechanism could be inappropriate antidiuretic hormone secretion (common in conditions of severe pain, perhaps an indirect effect of exercise).

Since the hyponatremia occurring with exercise develops acutely, it may be associated with several symptoms such as seizure, change in mental status, or coma. When these symptoms occur, emergency treatment with hypertonic saline is necessary. In milder cases, intravenous sodium chloride is usually adequate.

Prevention. Hyponatremia can only develop when water ingestion exceeds insensible losses. As long as water intake is less than or equal to insensible losses via sweat, serum sodium concentration should not change significantly. Therefore, replacement fluids should be consumed in quantities that do not greatly exceed losses. If volume losses were replaced with hypotonic electrolyte-containing fluids, the small risk of hyponatremia would be nearly eliminated.

Lactic Acidosis

Lactic acidosis may occur after exercise in association with heatstroke or as an isolated condition. The symptoms seen with lactic acidosis are identical to the symptom complex found in any form of metabolic acidosis. Dyspnea is common as the patient hyperventilates to compensate for the acidosis. Neurologic symptoms occur only rarely, causing lethargy or frank coma. The most important presentation, but also one that is very rare with exercise, is that of cardiac arrest: arrhythmias are common when blood pH falls below 7.1.

The etiology of exercise induced lactic acidosis is related to an imbalance in the supply and demand for oxygen in muscle cells. The most efficient conversion of glucose to adenosine triphosphate (ATP) occurs by complete passage from glycolysis through the Krebs cycle and on through oxidative phosphorylation. When oxygen is not available, as in muscle cells during intense exercise, pyruvate, the end product of glycolysis, is converted into lactic acid. The acid produced is rapidly buffered in plasma, and the lactate is converted back into pyruvate in the liver and kidneys.

During intense exercise, lactate levels in the blood can rapidly increase twentyfold.[15] In extreme exercise, lactic acid is produced in such great quantities that it cannot be effectively buffered; systemic pH may transiently fall to as low as 6.8 under these conditions.[18] Lactate is rapidly metabolized by the liver during rest, correcting the systemic pH to normal. It is very rare for acidosis occurring during exercise to cause any clinically important problems in healthy individuals. When acidosis persists after rest, hepatic dysfunction, with resultant decreased metabolism of lactate, should be suspected.

Acute Renal Failure

An uncommon but quite dangerous complication of exercise is acute renal failure. The pathogenic mechanisms relate to the loss of fluids and electrolytes as discussed above. This condition should be suspected in a patient who has recently performed strenuous exercise, has decreased urine output, or has symptoms of uremia.

The mechanism of acute renal failure in this setting probably involves the combined effects of hypovolemia and rhabdomyolysis. The hypovolemia leads to renal ischemia and decreased flow through the renal tubules. This results in increased myoglobin concentrations within the tubules, which predispose the patient to acute tubular necrosis by mechanisms that are not completely understood.

Significant rhabdomyolysis as a result of exercise is uncommon. However, it is important that it be recognized early by the clinician, because immediate prophylactic measures may help prevent acute renal failure. Signs and symptoms of rhabdomyolysis include diffuse muscle pain, muscle swelling, hyperkalemia, hyperphosphatemia, hypocalcemia, and hyperuricemia. The patient's urine will be red, and the urinalysis will be positive for heme by dipstick, but examination of the sediment will not reveal red blood cells. Only demonstration of elevated serum levels of muscle enzymes, such a creatine phosphokinase, lactate dehydrogenase, and serum glutamic-oxaloacetic transaminase, will establish the diagnosis conclusively.

Exercise-associated rhabdomyolysis is usually related to severe muscle exertion. The astute clinician should keep in mine, however, other predisposing factors, such as hypophosphatemia, hypokalemia, alcohol or cocaine ingestion, certain medications such as lovastatin and gem-fibrozil, and viral infection.

When the diagnosis of rhabdomyolysis has been established, preventing renal failure is crucial. The primary measure should be administration of enough intravenous fluid to maintain urine flow rate of at least 200 mL/hr. Forced alkalization of the urine with infusions of bicarbonate and mannitol to maintain the urine pH greater than 6.5 may be effective.

The risk of developing acute renal failure related to exercise appears to be greater after prolonged, strenuous exercise on hot days in underconditioned athletes.[20,21] Among military recruits, acute renal failure has been noted in the summer during the first few weeks of basic training.[21]

Prevention. Besides adequate replacement of fluids and electrolytes during and after exercise, other steps can be taken to avoid acute renal failure. First, people embarking on a new exercise program should increase the intensity of their workouts gradually. Second, athletes traveling to a climate hotter than that in which they usually exercise must realize that they are not fully heat acclimated. They must, therefore, initially temper the intensity of their exercise. Finally, on very hot days, anyone, regardless of their level of conditioning or heat acclimation, must exercise with care.

Abnormalities Found on Urinalysis

In the condition described above, the extreme metabolic stress of exercise on muscle leads indirectly to renal dysfunction. In contrast, direct and hormonal effects on the kidneys are probably more important in the pathogenesis of hematuria and proteinuria.

Hematuria. Either gross or microscopic hematuria can occur with exercise. The prevalence of microscopic hematuria depends on the type of activity, ranging from 55% in football to 80% in swimming or track and field.[22] One should keep in mind that red urine does not necessarily indicate hematuria, because myoglobinuria and hemoglonuria also cause the urine to appear red. Red blood cells in the urine sediment confirm the diagnosis of hematuria.

The mechanism of hematuria occurring after exercise varies. With contact sports such as football, direct trauma to the kidneys is likely an important factor. With long distance running, it has been suggested that trauma to the bladder from the repeated up and down motion might be important.[23] Others[24] have found red cell casts and dysmorphic red blood cells in the urinary sediment, suggesting bleeding of glomerular origin. It is not known how exercise might cause glomerular bleeding.

Hematuria is a common presenting problem in clinical practice. When hematuria is due to exercise, the prognosis is excellent; the condition usually resolves within 48 hours. It is crucial, therefore, to inquire about recent strenuous exercise in the initial evaluation. When exercise is suspected to be the cause of the hematuria, a urinalysis should be repeated 48 to 72 hours after a period of rest; if the urine is free of blood at that point, no further workup is indicated. If the hematuria resists, a full evaluation as indicated by the patient's age should be initiated.

Proteinuria. Urinary excretion of protein is increased after strenuous exercise in 70% to 80% of subjects.[25] Dipstick readings for protein are found to be in the range of 2+ to 3+, with maximal excretion in the 30 minutes following exercise.[22] Twenty-four hour collections of urine after exercise contain up to 300 mg of protein (normal<150mg).

The mechanism of proteinuria occurring after exercise is probably related to a decrease in plasma volume as discussed earlier. This leads to increased renin and angiotesin II secretion,[26] with a resultant increase in the glomerular filtration fraction. With the increased filtration of water, the contraction of plasma proteins in glomerular capillaries rises, causing increased movement of protein across the glomerular basement membrane by means of diffusion and convection.[27]

Because proteinuria is a common clinical finding, the initial evaluation of any patient with proteinuria begins with a consideration of factors that may cause transient proteinuria, such as exercise. (Other causes include congestive heart failure and fever.) As with hematuria, proteinuria should always abate within 48 hours, and the prognosis in such cases is excellent. When proteinuria persists beyond this point, and evaluation for an underlying renal disease is necessary.

An Exercise in Prevention

When an active patient appears to have a renal illness or injury, it's important to elicit a history of recent strenuous exercise. This information is essential for proper diagnosis and expedient treatment. In addition, sports physicians need to stress to their patients the role of adequate fluid and volume replacement in both preventing and immediately correcting renal and electrolyte complications.

References

1. Better O.S.: Impaired fluid and electrolyte balance in hot climates. Kidney Intl Suppl. 1987, **21**:S97-S101.
2. Costill D.L., Cote R. Fink W.: Muscle water and electrolytes following varied levels of dehydration in man. *J. Appl Physiol* 1976:**40**(1):6-11.
3. Rowell L.B., Brengelmann G.L., Blackmon J.R., et al: Splanchnic blood flow and metabolism in heatstressed man. *J. Appl Physiol* 1968: **24**(4): 475-484.
4. Knochel J.P., Beisel W.R., Herndon E.G. Jr, et al: The renal, cardiovascular, hematologic and serum electrolyte abnormalities of heatstroke. *AM J Med* 1961: **30**:299-309.
5. Austin M.G., Berry J.W.: Observations on one hundred cases of heatstroke. *JAMA* 1956: **161**(16):1525-1529.
6. Coyle E.F., Montain S.J.: Benefits of fluid replacement with carbohydrate during exercise. *Med Sci Sports Exerc* 1992: **24** (9 suppl):S324-S330.
7. Noakes T.D.: Fluid replacment during exercise. *Exerc Sport Sci Rev* 1993: **21**:297-330.
8. Rosa RM, Silva P., Young JB, et al: Adrenergic modulation of extrarenal potassium disposal. *N Engl J Med* 1980:302(8):431-434.
9. Struthers A.D., Quigley C., Brown M.J.: Rapid changes in plasma potassium during a game of squash. *Clin Sci* 1988: **74**(4):397-401.
10. Rose L.I., Carroll D.R., Lowe S.L., et al: Serum electrolyte changes after marathon running. *J Appl Physiol* 1970: **29**(4):449-451.
11. Coester N., Elliott J.C., Luft U.C.: Plasma electrolytes, pH, and EGG during and after exhaustive exercise. *J Appl Physiol* 1973: **34**(5):677-682.
12. Ledingham I.M., MacVicar S., Watt I., et al: Early resuscitation after marathon collapse, letter, *Lancet* 1982: **2**(8307):1096-1097.
13. Clausen T., Wang P., Orskov H., et al: Hyperkalemic periodic paralysis: relationships between changes in plasma water, electrolytes, insulin and catecholamines during attacks. *Scand J Clin Lab Invest* 1980: **40**(3):211-220.
14. Knochel J.P., Blanchley J.D., Johnson J.H., et al: Muscle cell electrical hyperpolarization and reduced exercise hyperkalemia in physically conditioned dogs. *J Clin Invest* 1985: **75**(2):740-745.
15. Lindinger M.I., Heigenhauser G.I. , McKelvie R.S., et al: Blood ion regulation during repeated maximal exercise and recovery in humans. *Am J Physiol* 1992: **262**(1 pt 2):R126-R136.
16. Lim M., Linton R.A., Wolff C.B., et al: Propranolol, exercise, and arterial plasma potassium, letter, *Lancet* 1981: **2**(8246):591.
17. Irving RA, Noakes TD, Buck R, et al: Evaluation of renal function and fluid homeostasis during recovery from exercise-induced hyponatremia. *J Appl Physiol* 1991:70(1):342-348.
18. Osnes J.P., Hermansen L.: Acid-base balance

after maximal exercise of short duration. *J Appl Physiol* 1972: **32**(1):59-63.

19. Better O.S., Stein J.H.: Early management of shock and prophylaxis of acute renal failure in traumatic rhabdomyolysis, *N Engl J Med* 1990: **322**(12):825-829.

20. Schrier R.W., Henderson H.S., Tisher C.C., et al: Nephropathy associated with heat and stress and exercise. *Ann Intern Med* 1967: **67**(2):356-376.

21. Vertel R.M., Knochel J.P.: Acute renal failure due to heat injury: an analysis of ten cases associated with a high incidence of myoglobinuria. *Am J Med* 1967: **43**(3):435-451.

22. Cianflocco A.J.: Renal complications of exercise. *Clin Sports Med* 1992: **11**(2):437-451.

23. Kallmeyer J.C., Miller N.M.: Urinary changes in ultra long-distance marathon runners. *Nephron* 1993: **64**(1): 119-121.

24. Fassett R.G., Owen J.E., Fairley J., et al: Urinary red-cell morphology during exercise. *Br Med J* (Clin Res) 1982: **285**(6353):1455-1457.

25. Alyea E.P., Parish H.H. Jr.: Renal response to exercise-urinary findings. *JAMA* 1958 **167**(7):807-813.

26. Costill D.L/, Branam G/, Fink W/, et al: Exercise induced sodium conservation: changes in plasma renin and aldosterone. *Med Sci Sports* 1976: **8**(4):209-213.

27. Goldszer RC, Siegle AJ: Renal abnormalities during exercise, in Strauss RH (ed): Sports Medicine, ed 2. Philadelphia, WB Saunders Co., 1991, p. 156. ❏

Questions

1. What are some of the ramifications to the body that occur routinely with physical activity?

2. How much sweat can be lost from the body during strenuous exercise on a hot day?

3. What are the primary ingredients found in the sports drinks recommended to replace water and electrolyte losses?

Answers are at the back of the book.

Section Six

Health and the Environment

37 *Coronary heart disease kills more adults worldwide than any other disease. For this reason, researchers are drawing their attention to the* Chlamydia pneumoniae *bacterium to see if it really can trigger coronary heart disease. Experiments with the laboratory strain called TWAR are showing high levels of its antibodies in the blood of people with coronary heart disease. Not everyone is a believer, however, as cardiovascular researchers can't agree whether an infectious agent like chlamydia is capable of playing a major role at all in the disease. Meanwhile, as the results of long-term studies unfold, the "TWAR wars" among the research community continues to rage.*

Can You Catch a Heart Attack?

Phillida Brown

New Scientist, June 8, 1996

Ten years ago, most people scoffed at the idea that a bacterium caused stomach ulcers. But linking chlamydia to coronary heart disease is seen as an even bigger heresy.

It's spread by coughs and sneezes. And by the time you're 20, there's a fifty-fifty chance that you're carrying the infection. But could *Chlamydia pneumoniae* really trigger coronary heart disease, that narrowing of the arteries which suffocates the heart muscle and kills more adults worldwide than any other disease? The theory is fueling one of the fiercest controversies ever seen in the multimillion-dollar industry of cardiovascular research.

The bacterium—often call TWAR after the original laboratory strain—is the prime suspect because its fingerprints keep appearing at the scene of the crime. People with coronary heart disease are likely to have high levels of antibodies to TWAR, suggesting that they have been persistently or repeatedly infected with it. TWAR's DNA and proteins keep showing up in atheroma, the fatty, diseased tissue that blocks the coronary arteries. And now, in as yet unpublished research, the bacterium itself has turned up alive and kicking, forcing the skeptics to sit up and take notice.

Some chlamydia experts have even come up with a plausible way of explaining how the common airborne bacterium that enters the body viz the lungs could travel to the blood vessels of the heart and wreak havoc. While others claim that *C. pnuemoniae* may explain some of the great puzzles in the pattern of the West's epidemic of heart disease.

The researchers are divided into three camps. The first camp suspects that the bacterium is fueling the process of atherosclerosis that damages tissue and narrows the arteries; the second, think it is in the arteries, but merely as an innocent bystander; and the third, doubt whether it is there at all. "It's TWAR wars," says Mark Enzler, a researcher at the Mayo Clinic in Rochester, Minnesota. But everyone agrees on one thing: the question must be answered soon. Coronary heart disease is so common that, even if a bacterium is playing only a bit part, thousands of people could die for want of a simple antibiotic. In Britain alone, for example, almost 170,000 people died of coronary heart disease in 1993. If *C. pneumoniae* triggered even 5 percent of those deaths as many as 8500 people might still be alive.

The TWAR wars will intensify this September when researchers gather in Vienna for the first major chlamydia meeting in four years. The factions refer to each other as believers, disbelievers and agnostics, and the believers see themselves as David's pitted against the Goliaths of the establishment.

There's a sense of *déjà vu* in all this. It is only in the past five years that another bacterium, *Helicobacter pylori*, has been officially recognised as the prime cause of stomach ulcers. For decades, the establishment view had been that, like heart disease, ulcers were the price of an unhealthy lifestyle. People who had been lectured for years about avoiding stress and eating regular meals suddenly had their ulcers cured with antibiotics, often in as little as a week. Coronary heart disease is a lot more complicated, but the chlamydia believers are not slow to draw parallels.

Thomas Grayston, an epidemiologist at the University of Washington in Seattle, is so suspicious that *C. pneumoniae* is directly linked to atherosclerosis that he proposes to find out if antibiotics such as tetracycline or erythromycin will improve the survival rate of people with coronary heart disease. If heart attack patients who receive antibiotics as well as "clotbusters" survive longer and in greater numbers than those who receive only clotbusters, he says, the unbelievers will be forced to rethink. "With *Helicobacter*, nobody paid much attention until treatment trials showed [stomach ulcers] could be affected by antibiotics."

Separate and Different

It was Grayston and his colleagues who first identified *C. pneumoniae* and showed, in 1989, that it is a separate species form its relatives such as *C. trachomatis*, a sexually transmitted bacterium. Grayston now wonders whether a chlamydia link can help to explain why death rates from heart disease are failing

rapidly in most industrialised regions of the world. Since the mid-1960s, death rates have plummeted by 60 percent in North America, and by up to 40 percent in Western Europe. They have also fallen steeply in Australia. Japan's death rate from the disease, which was always much lower than the West's, has also dropped. Only in Easter Europe are the death rates from heart disease rising.

Improvements in lifestyle cannot easily explain the widespread retreat of coronary heart disease. While the proportion of men who smoke in the West has dropped from half to about a third, and average blood pressure and blood cholesterol levels have fallen, obesity and diabetes—both implicated in heart disease—have become more common. Grayston says that all the improvements together account for "at most half" of the decline in death rates. "The reasons for the rest of the decrease are not known."

He points out that broad-spectrum antibiotics such as tetracycline which kill chlamydia, were introduced at about the time death rates started to fall in the US. If TWAR is playing a role in heart disease, then the widespread use of such antibiotics might be contributing to the decline in the death rate, he says.

But that's just naive, say other, more mainstream epidemiologists. The decline coincided not just with antibiotic use but also with "TV antennas, increased terrorist activity and the women's movement," scoffs William Kannel, and epidemiologist at Boston University and former director of the Framingham Study, the massive investigation that first identified the now familiar risk factors for coronary heart disease such as high blood pressure and cholesterol.

As for the idea that only half of the reduction in death rates from heart disease can be explained, the disbelievers argue that this view is simplistic. They say it doesn't take account of the fact that studies usually underestimate the impact of the known risk factors and fail to measure the ways in which these risk factors interact.

What is more, there is another explanation for the drop in deaths from heart disease that some experts prefer. This says that coronary heart disease is caused by malnutrition in the fetus and in early infancy, and the fall in death rates since the 1960s reflects improvements in mothers' nutrition earlier this century.

But not everyone thinks Grayston naive. Mike Rayner at the University of Oxford, a "believer" who studies changes in the incidence of heart disease, argues that as all attempts to explain trends in cardiovascular disease rely on large doses of speculation "any explanation of the trends is almost as good as any other." And with that caveat in mind, he points out that the rise of heart disease in the early 20th century and its rapid decline now follows the "classic" pattern of an epidemic, even if over an unusually long period of time. "So it is quite possible that infectious disease is playing a role," he says.

Searching for Clues

The controversial trial that first implicated *C. pneumoniae* in heart attacks began at the end of the 1980s when Finnish researchers led by Pekka Saikku, now at the National Public Health Institute in Oulu, found that people with coronary heart disease are more likely to have high levels of antibodies to the bacterium than healthy people. During the 1990s, the evidence steadily mounted. Using the polymerase chain reaction, Grayston and his colleagues tracked down *C. pneumoniae's* DNA in samples of atheroma taken from people with coronary heart disease. They also picked out the bug's proteins in the atheroma, using fluorescent-labeled monoclonal antibodies that bind to them. All told, about 60 percent of the diseased arteries has some trace of the microorganism, compared to none in the 31 samples of healthy coronary artery. Soon, other laboratories—in Finland, Italy, Britain, Japan and the US—were finding traces of *C. pneumoniae* in diseased arteries too. And just last week a team led by Joseph Muhlestein at the University of Utah School of Medicine reported in the *Journal of the American College of Cardiology* that they had found *C. pneumoniae* protein in 79 percent of the atheromas from 90 patients.

Getting Warmer

It was last year, however, that the chlamydia trial really heated up. James Summersgill and his col-leagues at the University of Louisville, Kentucky, told their colleagues at a meeting on infectious diseases in San Francisco that they had managed to grow *C. pneumoniae* from a diseased coronary artery in a man about to have a heart transplant. For the first time there was solid evidence in diseased arteries of the bacterium itself—not just its molecular fingerprints.

Researchers in two other laboratories at the State University of New York, in Brooklyn, and the Providence Medical Center, in Southfield, Michigan, were able to confirm the key finding. While no one is claiming that the results prove chlamydia's guilt beyond a shadow of a doubt, "this is very convincing evidence that chlamydia is there," says Summersgill.

"I was a disbeliever until I saw this," says Charlotte Gaydos, who heads a chlamydia lab at the Johns Hopkins University in Baltimore. " I don't think yet we have irrefutable evidence, but I think the evidence is growing." Summersgill admits, though, that more studies are needed, not least because none of the labs managed to grow live bacteria from samples they turned up DNA and protein evidence of the bacteria.

Meanwhile the disbelievers are standing their ground. Ironically, one of the most vocal of their number, Margaret Hammerschlag, is the head of the lab at the State University of New York which confirmed Summersgill's finding. Hammerschlag now dismisses that result as little more than a fluke because her own team's new studies contradict the findings. In the April issue of the *Journal of Infectious Diseases*, the Hammerschlag team reported that they had failed to culture *C. pneumoniae* form 58 samples of atheroma from New York patients, and that only one sample showed traces of the bacteria by any of the other methods. "How do you reconcile our big fat zeros with similar studies that find up to half of the samples positive?" she demands. (Not that Hammerschlag necessarily thinks *C. pneumoniae* is totally innocent. She is also investigating a link between the bacterium and asthma.)

Like Hammerschlag, Enzler has searched in vain for TWAR in a set of atheroma samples. Unlike her, however, he thinks the bacterium is probably there if not as often as Grayston's lab claims—it is, after all, notoriously difficult to culture. Others suggest the Grayston labs' bacterial sightings may be due to contamination—the bane of all labs using PCR. But Grayston is unshakable: "I am confident that our positive results are not false," he says.

Whether *C. pneumoniae* turns out to be in most diseased arteries or just a few, a key question must still be answered: is *Chlamydia*, the commonplace microbe that may be spread in a cough or a sneeze, merely guilty of being in the wrong place at the wrong time or does it real injure the heart's blood vessels? Now, interested parties like Gaydos have even come up with plausible theory for how *C. pneumoniae* could do that damage: here's how.

Free Ride

Scientists studying atherosclerosis know that the build-up of atheroma is not just a crude furring-up, like sludge in a pipe, but inflammation and thickening due to an immune defense mechanism gone wrong. The trouble starts when the endothelium, the tissue lining the artery, is irritated by low-density lipoproteins (the so-called bad cholesterol), tobacco components and other substances. Immune cells called macrophages rush to the site of the injury, where they become bloated with lipoproteins. Next, they are drawn out of the blood and accumulate in the vessel wall, forming the fatty streaks that are the first signs of damage. As the damage grows, the inflammation worsens, more macrophages are attracted to the scene and the cells lining the blood vessel proliferate and grow fibrous. According to the theory, an infection of *C. pnuemoniae* in a blood vessel could help to fuel that process, or even start the whole thing off.

Inflammation and chronic infection are hallmarks of the chlamydia family. For instance, *C. trachomatis* inflames and scars the fallopian tubes. Put crudely, *C. pneumoniae* may be inflicting similar damage to the artery wall. For that to happen, however, it has to get from the lungs to the coronary arteries.

The crux of the theory, says Gaydos, is that *C. pneumoniae* hitches a ride on macrophages that pass through the lung's alveoli, enter the bloodstream, and then get pulled into the vessel wall at the site of an injury. Macrophages, after all, are all-purpose scavengers that eat not only low-density lipoproteins but also pathogens. Once *C. pnuemoniae* gets into the vessel wall, the infection would cause more inflammation and pull more macrophages into the tissue, leading to a downward spiral of damage.

So appealing is the theory that believers, agnostics and even professed unbelievers like Hammerschlag, are all investing time and effort to test it in the laboratory. Hammerschlag's team is infecting mice that are naturally prone to atherosclerosis with chlamydia to see whether they develop the condition earlier than uninfected animals. Grayston's lab has begun similar experiments with rabbits. And Gaydos has managed to infect human endothelial cells with *C. pneumoniae* in the lab, while others have infected human macrophages. Most recently, Gaydos tested whether a human macrophage can carry the bacterium into the smooth muscle cells below the lining of the artery—which would lend strong support to the *C. pneumoniae*-heart disease hypothesis. She won't, however, divulge the results just yet: her colleagues will have to wait for the Vienna meeting.

The cardiovascular research establishment, meanwhile, remains largely unconvinced. For those who are prepared to entertain the chlamydia hypothesis such as Gaydos, this is no surprise. "The gastroenterologists were the last to believe the *Helicobacter* evidence," she says. So there is a long slog ahead: the final answer will come only though long-term studies in people and animals. In the meantime, Hammerschlag's advice is strictly practical: "Don't take your prophylactic erythromycin just yet." ❏

Questions

1. What made scientists think that *Chlamydia pneumoniae* may be involved in coronary heart disease?

2. What exactly is the theory behind *Chlamydia pneumoniae* and heart disease?

3. What is responsible for the decline in death rates from heart disease since the mid-1960s?

Answers are at the back of the book.

38

Stress physiology is not a new area of modern medical science but it is one which is being given considerable attention as many people search for ways to reduce it. While reducing stress can help maintain health and fight disease, researchers are discovering other ways that stress could be beneficial. Once thought to be a suppresser of immune function, some stress may instead have the potential to enhance the immune response. Scientists discovered in test animals that leukocytes, a type of immune cell, were redistributed to the skin and other locations in the immune system by moderate amounts of stress. While this could pose as a short-term natural defense for the body, it still doesn't predict how chronic stress could affect the health of an individual over a lifetime.

Skin-Deep Stress

Mike May

***American Scientist,** May/June, 1996*

In the 1930s Hans Selye, a Canadian physiologist, discovered that stress can produce health problems. Over the years since then, studies on various species of animals have generated a fundamental principle of stress physiology: Stress suppresses immunity by inhibiting a variety of immune-system parameters, such as the number of white blood cells.

Now it turns out that stress may actually do something quite different. It may have the potential to enhance the immune response by redistributing—not destroying—white blood cells.

This turnabout has emerged largely from work performed by Firdaus Dhabhar, a doctoral candidate in neuroscience working in Bruce McEwen's laboratory at The Rockefeller University. Dhabhar constructed a series of experiments with a general purpose. "My interest in the effects of stress on immune function," he explains, "arose from an interest in understanding—using the techniques and principles of modern science—the mechanistic basis behind the age-old belief that matters of the mind

can affect the health of an individual." As a start, he examined how mild stress affects an animal's white blood cells.

Dhabhar first produced mild stress in rats by confining them in close-fitting Plexiglas tubes for two hours. During and after the stress test, Dhabhar took samples of each rat's blood and determined the white-cell count and the level of corticosterone, a hormone released by the adrenal cortex during stress. Soon after the test started, the white cells in the blood plummeted and corticosterone climbed. In fact, the white-cell count dropped by nearly half after two hours of stress. Nevertheless, the count returned to normal in an hour or so after the end of the stress test.

The correlation between the increase in corticosterone and the decrease in white blood cells turned out to be a causal relationship. After removing the adrenal glands (which secrete stress hormones) from some rats, Dhabhar found only a small stress-related decrease in the blood's white cells. When those rats

Reprinted with permission from *American Scientist,* May/June 1996.

were given corticosterone, however, stress caused "normal" decreases in their white-cell counts. So adrenal hormones apparently mediate the interaction between stress and white-cell blood levels.

For the most part, other experiments had already shown that stress triggers the release of specific hormones and a decreased white-cell count. So Dhabhar went on to examine a more intriguing question: What happens to the white blood cells? Maybe stress reduces the turnover of white blood cells—the ratio of production to destruction—thereby decreasing the number of cells. On the other hand, he hypothesized, stress might redistribute white blood cells by directing them out of the blood and into other body compartments, which would reduce the number of *circulating* white cells. In the May 15, 1995, issue of *The Journal of Immunology* (154(10):5511-5527), Dhabhar and his coauthors wrote that the second explanation was more plausible: "It is unlikely that over 40 to 70% of the circulating leukocyte pool would be destroyed within 2 [hours] of exposure to mild stress, and completely replaced within 1 to 3 [hours] after the cessation of the stress." In addition, the investigators had measured the levels of lactate dehydrogenase, an indicator of cell damage, and it did not change significantly—meaning that white blood cells were probably not being destroyed.

In response to those results, Robert Sapolsky of Stanford University asked: "What does a change in some particular immune measure mean? That is often a particularly hard thing to figure out—whether it means anything about disease down the line—when it is done in a test tube, and when only a few things are measured. The strength of this study is that it is done in a real whole animal, in response to a real physiologic stimulus, and a lot of things are measured."

Dhabhar also wondered what his results meant. To find out, he and McEwen induced allergic contact sensitivity—like a poison-ivy reaction—in the skin of rats and compared the immune response in stressed and nonstressed rats. The stressed rats produced a faster, longer-lasting and more extreme inflammatory response than the nonstressed rats. Stress had enhanced, not suppressed, the immune response. In addition, Dhabhar and McEwen found more white blood cells in the skin of stressed rats, compared with nonstressed rats. This raised the possibility that stress might concentrate white cells in the skin and other locations in the immune system, including the lymph nodes and bone marrow.

"An important function of endocrine mediators released under conditions of acute stress may be to ensure that appropriate leukocytes are present in the right place and at the right time to respond to an immune challenge which might be initiated by the stress-inducing agent, such as an attack by a predator or invasion by a pathogen," Dhabhar suggested. "The modulation of immune-cell distribution by acute stress may be an adaptive response designed to enhance immune vigilance and increase the capacity of the immune system to respond to challenges in immune compartments, such as the skin, which serve as major defense barriers for the body."

What do Dhabhar's studies tell us about living in these stressful times? Should you forget about all the stress-reducing strategies, the ones intended to tone down any hostile tendencies in your personality? Should you seek out stress to build a power-packed immune system? Dhabhar wouldn't recommend it. "While moderate and manageable stress may have beneficial effects, chronic stress may be bad for you," he points out. "Moreover, it is important to recognize that the enhancement of immune function may be beneficial in certain cases—effects of vaccination, resistance to infections and cancer—but harmful in others—allergies, arthritis and autoimmune diseases." But Dhabhar believes that new insights on stress can only help. "The long-term aim of my studies," he said, "is to understand how one can efficiently harness the body's natural defenses to maintain health or to fight disease." ❑

Questions

1. How does stress affect the immune system?

2. Where do leukocytes concentrate in the body under acute stress?

3. What hormone is released from the adrenal glands during stress?

Answers are at the back of the book.

39

Have you ever thought about the chemicals you might be ingesting when you eat or drink? Chemicals such as DDT, PCBs and dioxin, whose uses have been restricted for years, still persist in the food chain and scientists believe that some of them can imitate natural hormones already in our bodies. These endocrine-disrupting chemicals, which the body cannot degrade, accumulate in the body fat and then migrate to target organs where they can derail reproductive, immune and neurological development. Environmentally relevant amounts of chemicals are being detected and changes are already evident in some wildlife populations. The human species really doesn't have the time to evolve the necessary defense mechanism to combat the hormone disrupters, so in the meantime, scientists must carefully identify which chemicals are the real culprits in the hormone charade.

Hormone Hell

Catherine Dold

Discover, September 1996

Industrial chemicals—from plastics to pesticides—paved the road to modern life. Now it appears that these same chemicals, by mimicking natural hormones, can wreak havoc in developing animals. And the road we once thought led to material heaven is heading somewhere else entirely.

Biologist Charles J. Henny reaches into a plastic bag and pulls out eight long, slender structures that look something like old chicken bones. He carefully places them side by side on his desk and then points out the obvious. "See how they get smaller and smaller," he says, waving a hand over the lineup. "It's right in correlation with the contaminants." The evidence, thus laid out on the gray metal desktop, seems clear. Otters living in the lower Columbia River area near Portland, Oregon, have a serious problem.

The thin sticks in question are not chicken bones but baculums, the bony part of a river otter's penis. Those on the left side of the lineup once belonged to otters number 28 and 29, "reference" animals that were taken from a less contaminated river habitat miles away from the Columbia. At nearly six grams each, they are significantly larger than the remaining six baculums, which were taken from Columbia River otters. These Columbia specimens average just 2.62 grams, with the smallest weighing a measly 1.92 grams. The otters' testes, says Henny, show a similar range in size, all the way down to one poor otter that didn't appear to have any testes at all.

To the naked eye, all eight animals had seemed healthy—all were just under a year old and about the same size, around 15 pounds. They had all been caught by fur trappers, who then froze and kept the skinned carcasses until Henny could collect them and bring them back to his office at the National Biological Service in Corvallis, Oregon. When he and a veterinary pathologist examined the otters, the only significant difference they detected—besides the weight and size of the reproductive organs—was in the levels of industrial chemicals and pesticides in the animals' livers. Time after time, when they analyzed the tissues for PCBs, heptachlor, mirex, or one of several dioxin-like compounds, the relationship was clear: the higher the concentration of chemicals, the smaller the reproductive organs.

"It was unbelievable to see those baculums line up the way they did," Henny remarks a short time later as he steers his pickup along the banks of the Columbia River. At river mile 119, a few miles east of downtown Portland and 119 miles inland from where the Columbia finally empties into the Pacific Ocean, he pulls off the highway. "This is near the famous spot where Lewis and Clark shot a condor," he says. Looking at the wild, wide Columbia River, whipped with rain under a steel gray sky, it's easy to imagine the legendary explorers scouting around the river and the dense forests, maybe even trapping a few otters themselves.

Looks can be deceiving. Although the Columbia doesn't much resemble eastern rivers that fairly scream pollution, with smoking factories lined up toe-to-toe on their banks, it is polluted nonetheless. Heavy metals, dioxins, furans, PCBs, DDT, and other pesticides are all there. Some came from local industry and farm runoff; some were probably transported on air from other parts of the globe. Some of the pollutants exceed allowable levels; some don't. For some chemicals, permissible levels have not been set. Most of the pollutants tend to accumulate in animal fat, and the otters, eating at the top of the local food chain, seem to be getting plenty.

• • •

Thirty-four years ago Rachel Carson's *Silent Spring* alerted the world to the dangers of pesticides. Chemicals such as DDT were recognized to be fatally toxic to some species and to cause widespread reproductive failure among others. Now scientists are finding that these same chemicals, at lower concentrations, can have an array of unexpected effects. Acting in the earliest stages of an animal's development, these chemicals are believed to play havoc with hormonal systems, leading to abnormal reproductive organs, skewed sex ratios, odd mating behavior, and animals that seem to be neither entirely female nor entirely male.

The Columbia River otters are obviously affected. And many other species—from alligators in Florida to beluga whales in the St. Lawrence River—are showing similar problems. It seems reasonable to wonder, then, if these chemicals pose a threat to humans also. After all, we are not so different from other animals, and some researchers think these same chemicals might well be pushing our sperm counts down and our cancer rates up. Added to this worrisome possibility is recent evidence that points to what could be an even more insidious effect; some of these endocrine-disrupting chemicals appear to be altering the behavior of children. And they seem to be doing it at relatively low levels—levels that many of us already carry in our bodies.

Hormones, for all their notoriety in shaping sexuality, are little more than the messengers of the endocrine system. Hormones released by the pituitary gland trigger the appropriate release of hormones elsewhere in the body, such as in the ovaries or adrenal glands. These hormones, in turn, travel to other parts of the body to tell cells what to do and when to do it. In a woman, for example, the hormone estrogen tells the uterus to get ready to receive a fertilized egg; adrenaline tells the heart to beat faster.

In the fetus, however, hormones do more than orchestrate activity. They perform complex developmental tasks, tasks that require precise dosage and exquisite timing. They tell tissues whether they should become female or male reproductive organs, nerve cells, muscle cells, or even eyelash cells. Hormones set off this differentiating process by binding to a specialized molecule—a receptor—on the surface or in the interior of a cell. The hormone-receptor complex then informs the cell's DNA which genes need to be turned on, and the genes, in turn, tell the cell which proteins and other substances it needs to make to take on the structure and function of the cell it's fated to be. Hormones are what tell the fetal cell what it will be when it grows up.

But what if chemical impostors interfere with these carefully articulated messages? Many researchers now believe that a small army of common chemicals can somehow imitate natural hormones, binding to receptors on fetal cells and scrambling the genetic instructions. By causing a cell to turn on the wrong gene, or effectively turn off the right one, or even turn up the "volume" of a gene, these mimics can derail an animal's development, permanently distorting its reproductive, immune, and neurological systems. There are more than 50 of these endocrine-disrupting chemicals, as they have come to be called, most of them ubiquitous in our environment. Some, such as DDT, alachlor, atrazine, chlordane, dieldrin, heptachlor, and mirex, are pesticides. Others, such as PCBs, endosulfan, bisphenol-A, dioxin, and heavy metals, are chemicals that have been used in, or created as by-products in, the manufacture of such everyday products as paper and plastics.

In the United States the use of some of these chemicals was restricted decades ago. Still, they persist in the food chain because of the way they accumulate in animal tissue: the chemicals, which the body cannot degrade, tend to lodge in fat; animals feeding at the top of the food chain usually have higher levels because they absorb the chemicals that have accumulated in their prey. Moreover, many of these chemical continue to be used in developing countries. The ultimate result is that these substances can be found nearly everywhere on Earth. Blackfooted albatross on Midway Island in the middle of the Pacific Ocean are contaminated with DDT, PCBs, and dioxin. Beluga whales in the St. Lawrence River have PCB levels so high they must be treated as hazardous waste when they die. Even marine mammals and people living in remote Arctic regions carry DDT, PCBs, dioxins, and other chemicals, transported around the world in the atmosphere. "We have no evidence that there are any populations that don't have these chemicals—fish, wildlife, or people," says Linda Birnbaum, director of experimental toxicology at the Environmental Protection Agency. Ninety to ninety-five percent of the suspected endocrine-disrupting chemicals that we absorb, she adds, are thought to come from the food and water we consume.

Fetuses—of any species—are particularly sensitive to exposure. When a pregnant female breaks down her fat reserves, the chemicals migrate into the fetus, accumulating at concentrations many times greater than daily adult exposures. Once there, they may unleash far more powerful effects than in an adult, some of which may not become apparent until sexual maturity.

One series abnormality after another has been reported in wildlife that have been exposed to a highly contaminated environment. Alligators in Lake Apopka, Florida—the site of a cleaned-up toxic spill—have tiny penises; male fish in polluted English rivers are producing a protein normally found only in fish eggs; beluga whales seem to be having fertility problems. In the case of the Columbia River otters, scientists found a clear dose-response curve— as the levels of contaminants increased, the size of the reproductive organs decreased. In most wildlife cases, however, investigators have simply noted gross reproductive abnormalities associated with high levels of chemicals. It has not been possible to show dose-dependent relationships, nor, because these are wild populations and not controlled experiments, has it been possible to show definitive cause-and-effect relationships. Even in the case of the otters, it is impossible to say which of the contaminants are the problem. The researchers know they have a lot to learn.

• • •

Laboratory studies are starting to fill in the gaps. Controlled experiments have shown, for example, that PCBs applied at just the right time during development can change male turtles and alligators into females or "intersex" individuals. Exposing male gull embryos to DDT can cause them to develop ovarian tissue. Giving tiny amounts of dioxin to rats before birth can sharply reduce sperm generation, "feminize" male mating behavior, and decrease the size of male sex organs. Most of these chemicals were thought to act through the estrogen receptor, but recent studies have turned up other routes. A derivative of DDT known as DDE was recently found to interfere with normal male development by binding to receptors for androgens— that is, "male" hormones, such as testosterone— and blocking their effects. Other chemicals have been found to attach themselves to "orphan" receptors, molecules whose intended function is unknown. To make matters worse, animal studies have recently turned up evidence of an awful chemical synergy; evidently, two hormone-mimicking chemicals can exert far more powerful effects than either chemical alone.

Still, much of the field remains a mystery. "We don't really understand why these chemicals are capable of mimicking hormones," says Frederick vom Saal, a biologist at the University of Missouri who studies the effects of estrogenic chemicals on mice. "They don't look anything like estradiol"—the most potent form of natural estrogen. What is not mysterious, though, is the potential for damage across a wide range of species. Estradiol, vom Saal explains, is "the same estrogen that is present in the body of a fish, frog, reptile, human, or bird. It hasn't changed in the course of 300 million years of evolution. The receptor hasn't changed, either." Other hormonal systems have not been as thoroughly studied, but they, too, probably evolved from some common ancestor and are at least similar among vertebrates today.

All of which leads vom Saal and others to ponder, "Why should humans respond any differently to endocrine-disrupting chemicals?"

"If you look at the developing embryo right after conception, whether it's a rat or a human or an alligator, they are all very similar," says Theo Colborn, a senior scientist with the World Wildlife Fund who has taken the lead in publicizing the growing body of evidence on endocrine disrupters. "It is during these stages of development, before specialization, they are all vulnerable."

Critics of the endocrine-disrupting theories say that synthetic chemicals are "weak" and not nearly as potent as natural estrogen and thus not likely to produce ill effects in humans. "Arnold Schwarzenegger is weak relative to Superman," counters vom Saal. The real question, he says, is whether these chemicals are present in concentrations high enough to elicit a response.

"Human cells or rodent cells respond to estrogen at approximately one ten-trillionth of a gram per milliliter of blood," says vom Saal. "What if a synthetic chemical is 10,000 times less potent than that?

Then the prediction would be that a chemical present at only one part per billion could exert a biological response. It turns out that a can of peas contains as much as 30 micrograms of bisphenol-A." (This compound, a powerful estrogen mimicker, is found in the plastic coating in cans.) "That is 30 parts per million, 300 million times higher than the natural action of estradiol. The people who are running around saying these are weak chemicals don't tell you that."

In his own laboratory, vom Saal is looking at the effects of what he calls environmentally relevant amounts of chemicals. Rather than seeing how how much exposure will kill an animal, he says, "we ask how the system functions normally and how much of this chemical would be required to cause problems. Then we look at the literature and see how much humans are eating. Are the amounts that induce changes in animals relevant to what is seen in the environment? The answer is often yes."

Ethically, of course, it is impossible to do controlled dosing experiments on humans. But tragically, there are cases of accidental contamination that provide evidence of effects in humans. In Japan in 1968, and again in Taiwan in 1979, women ingested rice oil that was contaminated with PCBs. The children born to those women have suffered from physical and mental developmental delays, behavioral problems including hypoactivity and hyperactivity, abnormally small penises, and IQ scores five points below average.

The clearest evidence of endocrine disruption in humans, however, comes not from accidental exposure but from a reportedly "safe" synthetic estrogen that doctors prescribed to as many as 5 million pregnant women from 1945 to 1971. The drug, diethylstilbestrol (DES), was thought to prevent miscarriage. It is now recognized as an endocrine disrupter that can distort fetal development.

"We have seen all kinds of structural changes in the vagina, cervix, and uterus," says Raymond H. Kaufman of the Baylor College of Medicine, who has studied adult women who were exposed to DES in the womb. These women are also at risk for an uncommon cancer of the vagina and cervix, some immune system disorders, ectopic pregnancy, and premature birth. DES-exposed men have a slightly higher risk for some genital abnormalities and decreased sperm counts. Experimental studies of rats and other animals exposed prenatally to DES have found similar abnormalities.

• • •

The question, of course, is whether hormone-disrupting chemicals now found in the environment can also produce such dramatic alterations in human sexual development, and many investigations into that possibility are under way. But even more worrisome are reports showing that the chemicals may already be producing subtle changes in memory and behavior in children exposed to them before birth.

The first studies of this type were launched more than 15 years ago. Joseph and Sandra Jacobson, husband-and-wife psychologists at Wayne State University in Michigan, decided to look at the babies born to women who had eaten trout and salmon caught in Lake Michigan. Fish from polluted lakes, rivers, and coastal waters are well-known source of PCBs and other contaminants; they soak up so many toxins, in fact, that some states warn women to avoid eating sport fish not only during pregnancy but at any time during their childbearing years. The Jacobsons asked several thousand new mothers about their fish-eating habits and eventually studied the children of more than 200 of them. What they found, says Joseph Jacobson, is "the clearest evidence yet that PCBs are causing neurobehavioral problems."

The Jacobsons analyzed the PCB levels found in the blood of each baby's umbilical cord, which gives an indication of prenatal exposure. At birth, they found, children who had higher exposures to PCBs had smaller heads and lower weights. At seven months, they tested the babies for cognitive function by showing them two identical photos for about 20 seconds. One of the photos was then paired with a new photo and shown to the baby again. The normal response for an infant is to spend more time looking at the new picture, indicating that it recognizes the familiar one. The babies who had the highest exposure to PCBs, however, spent as much time looking at the old photo as the new one, suggesting either deficits in short-term memory or attention problems.

When the children were four years old, they were given a battery of cognitive tests. Again the highly exposed children showed memory impairments, this time in tests that asked them to recall progressively longer strings of words and numbers. The differences in scores between unexposed and the highest-exposed children, says Joseph Jacobson, "would be like ten points on an IQ test. We're not seeing mental retardation, but we are seeing that the children are just not doing as well." Jacobson suspects these problems may affect the children's ability to master reading and arithmetic skills.

These children were not living next to a toxic waste dump, nor had their mothers eaten PCB-laden fish every day during pregnancy. Their exposure to PCBs, while high, is still considered to be within the range of normal background exposure levels, says Jacobson. Other possible causes, such as lead exposure or the mother's intake of tobacco or alcohol, were ruled out.

Two other studies of children have found similar problems. In the Netherlands, researchers found that 18-month-old children born to women who carried relatively high but still "normal" levels of PCBs were more likely to be neurologically "nonoptimal." In fact, the higher their exposure to PCBs, the lower their neurological scores. In that study the mothers had eaten normal diets; they got their contaminants through their food and probably from water and air. Meanwhile, back in the United States, researchers at the State University of New York at Oswego have found that babies born to "high fish eaters"—women who in their lifetimes had eaten at least 40 pounds of fish from Lake Ontario—tested worse on several scales than did babies born to "low fish eaters" and "non-fish eaters." Fish in Lake Ontario are highly contaminated with PCBs, dioxin, hexachlorobenzene, DDE, mirex, and other chemicals.

"We looked at the kind of stuff a pediatrician assesses in a newborn," explains Edward Lonky, a developmental psychologist at the university. "One of our main findings was in regard to habituation, which is a measure of neurological intactness. It's one of the tests we use to assess fetal alcohol babies, crack cocaine babies, and babies exposed to environmental contaminants. You shine a light through the eyelids of a lightly sleeping newborn and you get a startle response. When the body settles down, you repeat the light. The startle response should habituate, or diminish, over repeated administrations." Normally an infant will show better habituation on the second day of testing. Lonky and his colleagues found that infants from the high-fish-eater group showed poor habituation responses as well as a greater number of abnormal reflexes and stress responses.

Lonky doesn't know what his findings might mean for the children as they grow. But he notes that rats fed fish from the lake have been shown to react abnormally strongly to fearful and frustrating events. His habituation studies suggest, but do not prove, that chemically exposed human infants may also overreact.

Chemical exposure appears to be the culprit in these behavioral studies, and though there is as yet no direct proof of that, researchers agree that the correlations are significant. But if chemicals are to blame, how might they be scrambling messages in the brain? One theory is that PCBs and dioxins are mimicking or blocking the action of thyroid hormones. These hormones help organize the fetal brain and promote the growth of neurons, the nerve cells that transmit information; severe disruptions in thyroid levels can even lead to mental retardation.

Whatever the cause, such effects may not be rare. "These were not people who were eating fish every day," stresses Linda Birnbaum of the EPA. "I believe the data suggest there are subtle changes going on in at least a portion of our population."

• • •

Not all scientists, of course, embrace the theory that synthetic chemicals are disrupting fetal development. Stephen Safe, a toxicologist at Texas A&M University, has often questioned some of the studies cited. The hypothesis that synthetic chemicals are mimicking hormones is "reasonable," he says, given some of the evidence in wildlife. But that doesn't mean they must have the same effects in humans.

"Our diet is chock-full of huge concentrations of natural endocrine disrupters," says Safe, adding that we consume only "trace amounts" of synthetic chemicals by comparison. Although Safe acknowledges that the synthetic chemicals tend to accumulate in the human body while natural chemicals are quickly metabolized and excreted, he argues that the natural chemicals still have an opportunity to take action. "How much is active? We really don't know. But we've got to take into account the fact that we take in huge quantities of endocrine disrupters."

Indeed, naturally occurring chemicals, such as phytoestrogens in plants, are known to disrupt animal reproduction. They have typically arisen through an ongoing evolutionary battle between plants and animals. If a plant happens to produce an estrogenic chemical that renders cows infertile, cow herds decline, and presumably populations of that now-uneaten plant flourish. Over time, though, cows that can somehow degrade the chemical will outbreed the infertile cows, and the plants will have to come up with a new defense. Humans, like cows and other animals, have evolved similar defenses against plant chemicals, say Louis Guillette, a reproductive biologist at the University of Florida who first reported the reproductive problems in alligators. We can usually make enough enzymes to degrade normal endocrine disrupters with little or no effect on bodily processes. But Gillette points out that the human species hasn't had time to evolve similar defense mechanisms against something cooked up in a test tube only 30 or 40 years ago.

While some researchers are trying to understand just how endocrine-disrupting chemicals act in animals, others are finding even more common chemicals to add to the growing list of estrogen impostors. Laboratory studies in England have shown that two phthalates, common chemicals in the manufacture of plastics, can mimic estrogen. These chemicals are used in many types of plastic food wrappings and may well be leaching into foods. Meanwhile a food coloring known as Red Dye No. 3, which lends color to hot dogs and a host of other common foods, has also be identified as an estrogen mimic and a possible carcinogen.

The EPA is paying close attention to hormone-disruption studies. "I take the wildlife findings very seriously, and I think they should serve as a warning for humans," says Lynn Goldman, the agency's assistant administrator for prevention, pesticides, and toxic substances. The EPA is now developing a national research strategy as well as new guidelines for screening chemicals. Unfortunately, notes Goldman, its efforts have already been hindered by congressional budget cuts. The National Academy of Sciences, for its part, recently convened a panel of scientists to assess what is known about endocrine disrupters.

Much of the credit for an increased awareness of the potential dangers of endocrine-disrupting chemicals must go to Theo Colborn, who has more than once been called the next Rachel Carson. "Carson focused on cancer," says vom Saal. "Theo Colborn has now shown us that there is a whole other set of information out there that was right in front of us and nobody saw it." Colborn downplays any comparisons. Carson struggled alone to get her message across, Colborn says, while she has had nothing but tremendous support from the scientific community. She does, however, hope that the studies of endocrine-disrupting chemicals will have as big an impact on the world as Silent Spring has.

Since World War II, production of synthetic chemicals has risen over 350-fold. No one is seriously proposing that all such chemicals be banned, of course. Some of them are invaluable for controlling pests and keeping water clean. As Colborn sees it, the answer to worrisome "environmental hormones" lies in screening all synthetic chemicals for developmental effects and creating chemicals that won't persist in the environment.

"You can't go back and rebuild a brain," Colborn says. "You can't go back and put more sperm cells in a male. But the beauty of this is that it's not the result of genetic damage. The blueprint for the normal individual is still there. What we have to do is make sure that while that blueprint is being followed, while the chemical messengers that tell this individual how to develop are doing their job, we're not introducing more chemicals into the environment of the womb." ❑

Questions

1. How do industrial chemicals pose a threat to humans?

2. Which animals are particularly sensitive to endocrine-disrupting chemicals?

3. Why are trace amounts of synthetic chemicals still toxic?

Answers are at the back of the book.

40

Exercise can't be touted enough as being beneficial to your health. But associated with prolonged endurance training, athletes suffer with disastrous injuries to bones and muscles from constant overuse. More dramatic physiological effects of intensive training are syndromes associated with the cardiovascular, respiratory, immune and reproductive systems. Problems such as these are becoming more prevalent as training regimes for athletes have grown more demanding. Fortunately, many of these dangers can be avoided through careful planning of training schedules and with more consideration for the physiology of high athletic performance.

Science Tracks Down the Training Dangers

Christopher Surridge

Nature, July 4, 1996

Exercise, as many of today's athletes are discovering, can be bad for you. Moderate exercise has a beneficial effect on health. But at the levels maintained by dedicated athletes it can have various harmful side-effects. Increasing demands on the body have begun to expose the subtle physiological costs of chronic exercise. Understanding the nature of these costs has become central to high athletic performance.

There are several reasons why the negative impact of excessive exercise has become increasingly prominent. Perhaps the most important is the fact that improvements in sporting achievements over recent decades have been the result of even more demanding training regimes. Club runners 30 years ago might have expected to cover 20-25 miles a week; but their modern counterparts may cover three times that distance.

Observations of Overtraining

In the past, knowledge of the dangers associated with high levels of exercise and how they can be overcome was derived chiefly from the practical experience of individual athletes and their trainers. More recently, an increasingly scientific approach to all aspects of sport means that a more systematic approach has been adopted to studying the physiological effects of intensive training.

As a result, it is now becoming apparent that, rather than presenting intractable limits on the levels of performance achievable by the human body, such phenomena are seen more as dangers that trap unwary athletes who fail to take account of physiology when drawing up their training schedules.

The most easily understood—and earliest recognized—conditions resulting from hard and continuous training are 'overuse' injuries due to the failure of muscles, bones or joints under a repeated, heavy load. For most adult athletes, such injuries, which include stress fractures of bones, strained muscles, snapped tendons and inflammations of joints (such as 'tennis elbow'), are inconvenient, but can be overcome with time.

For children, whose bones are still growing, the consequences can be more serious. Adult success in sports such as swimming and gymnastics can require hard training even before an athlete has reached teenage years. One study of young female gymnasts found that 11 percent already had fractures in the vertebrae of the lower back, four times the occurrence in the normal population. Fortunately, potentially disastrous injuries can be avoided through the careful planning of training schedules.

The most dramatic training-related syndromes concern the cardiovascular and respiratory systems. Sudden death due to heart failure is uncommon but not unknown. Indeed, the first marathon runner, Pheidippides, expired after his run from Marathon to Athens in 490 BC with the words "rejoice, we won"—although Dan Tunstall-Pedoe, medical director of the London marathon, has seen only four deaths during the race's sixteen-year history.

Such sudden deaths are usually the result of training exacerbating one of a number of congenital heart defects (although heart failure in professional cyclists with no such complicating factors has been known). The most common cause of sudden death in young athletes in the United States, for example, is due to a condition known as hypertrophic cardiomyopathy, in which the walls of the left ventricle of the heart become abnormally enlarged. The natural thickening of the heart muscle due to exercise further reduces the size of this chamber, resulting in a fatal heart attack when the heart is placed under extreme strain, as during competition.

Asthma, the chronic constriction of the bronchial airways, is another potentially fatal condition that is aggravated by training. Although no less serious than congenital heart conditions, its effects are less dramatic. It is also far more common.

Asthma and Other Effects

According to Mark Harries, Clinical Director of the British Olympic Medical Centre, between 10 and 15 percent of athletes, even at the highest level, report occasional symptoms of exercise-induced asthma. As with its rise in the general population, the factors that aggravate asthma in athletes are far from clear, for example cold, dry air induces asthma

more effectively than warm, moist air. But it can be easily treated by inhalation of bronchiodilaters, some of which are permitted for use by athletes.

Some of the deleterious effects of training for high-level competition are gender-specific. In particular, the reproductive physiology of women athletes is very sensitive to disturbance under extreme training regimes. Roger L. Wolman of the Royal National Orthopaedic Hospital in London has estimated that in sports such as cycling, running and rowing, more than half the women competing at international level have disturbed menstrual cycles (oligomenorrhea). Similarly, more than 60 percent of women gymnasts do not menstruate (amenorrhea), and delayed onset of menstruation is common.

Exercise Is Only One Factor

These dysfunctions may or may not be related to sports, and are not caused exclusively by exercise. Thus although an athlete's training is one of a number of contributing factors, pressures to maintain a specific physique —for example, light weight in rowing coxes or female gymnasts—can produce similar symptoms to those of sufferers from eating disorders such as anorexia.

The cause of athletic amenorrhea is unclear. One theory, of which Hans Kreitzer of the University of Limburg in the Netherlands, is a leading proponent, is that elevated levels of endorphins and cortisol, due to physical and emotional stress, act on the hypothalamus, interfering with its secretion of gonadotrophin-releasing hormone, and thus suppressing levels of gonadotrophins.

Most seriously, the low levels of oestrogen during amenorrhea result in a reduction in bone density which, Wolman and others suggest, can leave the athlete vulnerable to osteoporosis in later life. Reduced gonadotrophin levels have also been recorded in male marathon runners—though much less frequently than in women—and have been associated with low sperm counts.

There is much anecdotal evidence that athletes suffer from an abnormally high incidence of colds and influenza. Only recently, however, has it been conclusively demonstrated that extremes of exercise do indeed reduce the effectiveness of the immune system. In one study, for example, B.K. Peterson of the Copenhagen Muscle Research Centre showed that exercising to exhaustion reduces the level of T lymphocytes and natural killer cells for up to 20 hours in both athletes and untrained individuals.

Similarly, Laurel MacKinnon, of the University of Queensland, Australia, has reported that antibody levels in saliva are reduced by 70 percent in racing cyclists after two hours of hard pedaling. And studies by E. D. Bateman at the University of Cape Town, South Africa, have shown that the fastest runners in marathons and ultra-marathons have their lowest white blood cell counts and highest incidence of respiratory infection in the days following races.

As with amehorrhea, these immuno-depressive effects seem linked to increased levels of stress hormones. In addition, Richard Budgett, medical director of the British Olympic Association, suggests that some athletes engaged in heavy training are unable to sustain levels of glutamine, which is produced in muscle and is essential for the metabolism of lymphocytes. During exercise, muscle tissue uses up the glutamine, effectively starving the white blood cells.

Avoiding Chronic Fatigue

The suppression of the immune system due to prolonged endurance training does not seriously compromise long-term health but does have serious consequences for performance. The British athletes Diane Edwards and Seb Coe are only two of the many well-known examples of athletes who have failed to achieve their potential at international events due to infections—in both cases toxoplasmosis, usually found as an opportunistic infection of immuno-deficient individuals.

Intimately associated with a reduction in general health is the dramatic onset of chronic fatigue experienced by some athletes during periods of particularly intense training, or when suddenly changing training regimes. This so-called 'over-training syndrome' is thought by some to be related to postviral fatigue syndrome (myalgic encephalomyelitis, or 'yuppie flu') and is characterized by reduced performance, fatigue and depression.

Such a condition is particularly insidious for athletes as its treatment—light exercise only slowly increased up to normal training levels—runs counter to the common misconception that harder training always leads to better results. Its underlying causes are still very poorly understood.

A fall in the density of red cells in the blood may be involved, and can be countered by altitude training (or illegal blood doping). In a study of US college swimmers, about 10 percent of whom are chronically fatigued, W.P. Morgan, of the Kinesiology Department at the University of Wisconsin-Madison, showed that monitoring of athletes' mood and resting heart-rate helped to predict, and thus help prevent, its onset.

Usually, however, over-training syndrome strikes with little warning and can be overcome only by a period of two or three months of light training, which can ruin an athlete's preparation for a major event—such as the Olympics. ❑

Questions

1. Why have the negative aspects of excessive exercise become so important in recent times?

2. Which group of athletes is most susceptible to vertebral fractures?

3. How does excessive exercise suppress the immune system?

Answers are at the back of the book.

41 *Women have come a long way in sports with dramatic improvements in performance. But some physiologists are speculating that women could conceivably outperform men in certain competitive sports that require high levels of endurance. While men can boost the power of muscle for speed on the track, women can savor the long-distance race with more body fat stores. Having more fat means that they are less reliant on glycogen stores which is the first fuel to be depleted during athletic training. This gives women a distinct advantage in many endurance competitions such as long-distance swimming and marathons.*

Could Women Take a Lead Over Men in the Long Run?

Ayala Ochert

Nature, July 4, 1996

Are there sports in which women might one day out-perform men? As traditional views about the physical limitations of women change, the question no longer seems absurd. This is particularly so since, after a comparatively late start, women have made dramatic improvements in performance in almost every competitive sport over the past 50 years.

The fact that this rate of increase has been faster than that for men—whose competitive efforts started earlier—over this period means that, based on a simple forward projection of record achievements, women would overtake men early in the next century.

In practice, we are seeing the upward slope of a S-shaped curve that will eventually flatten off. "Whether the rates will slow abruptly or slowly, we don't know, but they will slow," says Brian Whipp, professor of physiology at St. George's Hospital in London, who has been closely studying the statistics.

But it is already possible to point to some sports—particularly those requiring high levels of endur-ance—in which women may have the potential to outperform men.

For most sports, particularly those highlighted in the Olympics, such a possibility is made highly un-likely by two major physiological differences be-tween the sexes: the proportion of body fat, and the mass of muscle. Speed on the track requires muscle power, and the male sex hormone testosterone in-creases muscle mass. Conversely, female sex hor-mones increase body fat—but fat, to an athlete, is just excess baggage.

But the factors that boost the limit performance in some less mainstream events, particularly those described as 'ultra-endurance,' are less well-defined. These include the longer triathlon races, which in-volve 3.8-km swim, followed by a 180-km cycle ride and then a 42-km run, as well as various other events (such as long-distance swimming) that are not included in the Olympics.

Women's potential in such events has come as something of a surprise. In the past, for example, they were considered to be physically incapable of running in conventional long-distance events, such as the marathon. Indeed, such a belief meant that the first women's Olympic marathon was not held until 1984.

The belief appeared to be given scientific backing by studies in which physiologists claimed to have demonstrated that women were less able to tolerate heat stress. But most such studies had a fundamental flaw, namely that they essentially compared the physiological characteristics of sedentary women to those of athletic men. The reason was simple: there were relatively few highly trained female athletes available for study.

About 20 years ago, however, in sharp contrast to prevailing ideas, some physiologists and others started to speculate that women might actually be better suited than men to endurance events. The reason, they suggested, was precisely the fact that women have a higher proportion of body fat.

In long-distance swimming, extra body fat is certainly a distinct advantage, providing both extra buoyancy and insulation. Penny Lee Dean, for example, held the world record amongst both men and women for swimming the English Channel for 16 years, up to 1994. And Lynn Cox successfully swam from Alaska to Russia—a feat previously deemed impossible by physiologists because of the enormous distance and extreme cold.

More controversial was the suggestion of a different type of advantage in running events. During endurance races, an athlete uses both fat and carbohydrate (in the form of glycogen) to fuel their muscles, and the ratio can be critical. By using more fat to provide this fuel, athletes can preserve their glycogen stores and so continue for longer.

Glycogen or Lipid Metabolism?

It is only when glycogen runs out that an athlete reaches the limits of his or her endurance. Women's extra fat, it was suggested, might mean that they are less reliant on glycogen metabolism, and can make better use of lipid-fat) metabolism.

Subsequent studies ruled out this claimed advantage of women, showing both sexes mobilize fat in much the same way, and the hypothesis has gener-

ally gone into disfavour. "The suggestion that women might be at an advantage because they have a higher body fat than men was absurd," says Ben Londeree of the department of health and exercise sciences at the University of Missouri. "It was like saying that fat men should perform better than leaner men."

But recent studies that concentrate on elite athletes are finding that there may still be gender differences in the way muscles are fueled in, for example, very long-distance running, David Speechly and his colleagues at the University of Witwatersand in South Africa, for example, studied male and female athletes who had recorded identical times for the marathon (42 km). When the same athletes ran 90 km, the women significantly outperformed the men.

In such events, "factors that we know are important in sprint events don't apply," says Lynn Fitzgerald, an immunologist and reader in sports medicine at Brunel University in west London, and herself a former world-record holder in several ultramarathon events.

Fitzgerald's own preliminary studies, carried out with a research student, Stuart Mackie, suggest that such gender-based differences in performance at very long distances are indeed linked to the respective ratios of fat and glycogen metabolism. "The physiological and psychological traits that are important for long distance events are not related to testosterone," she says.

Endurance training is well-known to have a 'glycogen-sparing' effect, raising the consumption of free fatty acids. But individuals with an inherent ability in endurance events may naturally use more fatty acids. Such individuals—who may turn out to be more often women than men—must work harder, as fat metabolism is less efficient than glycogen metabolism. But they are also able to keep going longer, because their glycogen stores take longer to deplete.

This possible gender difference could, according to Fitzgerald, explain the fact that, although only one in four competitors in ultra-endurance races are women, they are often found in the top 10 percent of those who finish. It remains unlikely that a woman would eventually win a race of this kind; their absolute physiological capacities are smaller than those of men. But such a result is no longer inconceivable. ❏

Questions

1. What role do hormones play in the proportions of body fat and muscle found between the sexes?

2. What physiological advantage do women have with higher proportions of body fat?

3. Which type of athletic training is well-known to have a glycogen-sparing effect in the body?

Answers are at the back of the book.

42

There's a new drug appearing in the gyms across the nation called insulin-like growth factor-1 (IGF-1) which promises to pump up muscles and enhance overall strength. Like growth hormone, IGF-1 is being used to stimulate muscle growth in healthy young powerlifters. Bodybuilders and other professional athletes are creating a black market for the naturally occurring peptide which is quickly replacing the position once held by steroids as the ultimate performance enhancer. Prized for being undetectable in the urine, these new drugs come at a very high price both economically and physically. If abused, the costly IGF-1 and growth hormone drugs could cause serious side effects such as facial nerve paralysis and fluid build-up around the brain. Without laws to control the sale of these drugs, unwary athletes could be competing with danger.

Beyond Steroids

Scott Veggeberg

New Scientist, January 13, 1996

Bodybuilders will stop at nothing in pursuit of the perfect physique—including taking a drug designed to treat motor neurone disease.

The quest for muscles with more power knows no bounds. It began in the 1950s with American and Russian weightlifters injecting themselves with steroids. By the 1980s, natural substances such as growth hormone were the unscrupulous athlete's choice. Now a new elixir rumored to confer tremendous stamina and muscle tone is throwing its weight around in the gym. It's called insulin-like growth factor-1 or IGF-1, and bodybuilders and powerlifters are frantic to get their hands on it. Despite its growing reputation as a potent anabolic agent, there are no laws controlling the sale of IGF-1 in the US or Europe. And no one is yet testing for it at athletic competitions.

"A couple of big name bodybuilders paved the way for IGF-1 use because they had tremendous results building muscle with it," says Peter Thorn, president of the US Powerlifting Federation. "IGF-1 is out there on the streets of America right now," says T.C. Luoma, editor-in-chief of *Muscle Media 2000*, a body-building magazine devoted to drug-enhanced strength.

"It's being sold out of the trunks of cars in Venice [California], and brown paper packages containing it are being discreetly handed out at southern California gyms . . . most body-builders 'know a friend who knows someone who has a cousin' who can get the stuff." And sports researchers think it likely that people in Europe are trying to pump up their muscles with IGF-1, too.

It's no accident that all this is happening just as biomedical research into IGF-1 is taking off. Thousands of papers on IGF-1 have appeared in medical journals over the past few years. Researchers hope that the peptide will eventually be used to treat a number of disorders, including motor neurone disease and dwarfism. But at the same time, it's clear that not all the IGF-1 produced by biotechnology companies is reaching the clinics. Manufacturers say they are having to institute elaborate safeguards to prevent this costly and scarce drug from being diverted to the black market.

IGF-1 can't be knocked together in a garage with a bit of glassware and a smattering of high-school chemistry. It's a naturally occurring peptide made up of a precisely structured chain of some 70 amino acids folded into a unique configuration. The peptide gets it name because about half of its amino-acid structure is identical to insulin and it can bind weakly to insulin receptors, as well as its own IGF-1 receptors on the surfaces of cells.

Like growth hormone, IGF-1 is essential for normal growth and development. Indeed the biological roles of the two are intimately intertwined. The traditional view is that growth hormone is produced by the pituitary gland, courses through the blood, and stimulates the liver to unleash IGF-1 peptides. These peptides then disperse and stimulate cells to divide and grow. But recent research suggests that growth hormone can also trigger a more localised release of IGF-1 in cartilage, muscle and other issues.

During normal development the effects of IGF-1 differ sharply from one tissue to another. In muscle cells the peptide stimulates the production of proteins and other cell components, while it mobilises fat for use as energy in adipose tissue. In lean tissue IGF-1 prevents insulin from transporting glucose across cell membranes. As a result the cells have to switch to burning off fat for energy. This effect of IGF-1 explains why some researchers are testing growth hormone, which stimulates the body to produce IGF-1, as a slimming drug.

Beefing up muscle cells while burning off fat sounds like a great recipe for athletic success. The catch is that not everything that stimulates tissue growth during development will work in full-grown healthy adults. Over the past decade it's become clear that injections of growth hormone or of IGF-1 can help children with deficiencies in these substances to grow to a normal height. Growth hormone can also help adults with hormone deficiency to lose fat and gain lean tissue. And according to controversial research by Daniel Rudman, professor of geriatrics and gerontology at the Veterans Affairs Medical Center in Milwaukee, a six-month course of growth hormone can increase lean tissue and reduce fat in the elderly.

Already, clinics have sprung up in Mexico, Switzerland and the US devoted to selling growth hormone therapy—not to people with hormone deficiencies but to healthy individuals seeking greater strength and sex appeal in their golden years. Does it follow that injections of growth hormone, or its handmaiden IGF-1, could benefit healthy young powerlifters?

Here hard facts begin to degenerate into the flab of anecdote and opinion. According to Luoma, users are calling IGF-1 "the most wonderful stuff in the world" for gaining muscle and losing fat. In *Muscle Media 2000*, an anonymous bodybuilder describes his experience with IGF-1: "I got ripped. It was really noticeable. I stay pretty lean most of the time, but this took it to a different level. Veins in my abs, that sort of thing."

Luoma admits such reports are of dubious value. "Most bodybuilders are using so many different drugs that it's almost impossible to tell whether the effects are from the IGF-1," he says.

Nevertheless, some scientists do say that IGF-1 could be used to stimulate muscle growth in healthy adults. David Clemmons, an endocrinologist at the University of North Carolina, began to take the idea seriously in 1993 when he put four men and three women between the ages of 22 and 47 on a diet low enough in calories to cause muscle and other tissues to break down, a process called catabolism. Clemmons found that he could protect the subjects' muscles from catabolism by giving them a combination of IGF-1 and growth hormone. And the combination was far more anabolic than either substance alone, he says.

Bodybuilders have apparently cottoned on to Clemmons's work. "Combining IGF-1 with GH is practised by some bodybuilders," says Luoma. Others seem to value IGF-1 more highly than growth hormone because it is the substance that does growth hormone's work in the body.

Manufacturers of IGF-1 are anxious to distance themselves from claims that it has profound anabolic effects in athletes. Few want to see the name of a medically promising product tarnished by abuse stories, as happened with growth hormone.

"There's no question that IGF-1 works brilliantly in rats," says John Ballard, managing director of GroPep Proprietary in Adelaide, an Australian-based manufacturer of IGF-1. "It stimulates their growth, it reduces any wasting conditions from diabetes, kidney failure or removal of parts of the gut." The trouble is, it's not very specific. When IGF-1 is injected into adult rats, non-muscle tissues also grow, such as in the thymus, spleen and kidneys. That, says Ballard, means IGF-1 is unlikely to have a selective anabolic effect on human muscles.

Black Market
And just because IGF-1 slows down the progression of motor neurone disease doesn't mean it works by directly stimulating muscle growth. More important here could be the drug's ability to make new nerve fibres grow, thus renewing connections to muscle cells. Besides, the scarcity and expense of IGF-1 on the black market would make it hard for most people to obtain significant amounts. A single vial containing about 0.1 milligrams of IGF-1 has a black market price tag of $600. It would take 10 or more such vials per day on a continuous basis to produce a biological effect, says Ballard.

However, some sports authorities, such as USPF's Thorn, believe that there are well-paid, professional athletes, not to mention nations hungry for Olympic success, who could support such an expensive habit. And the fact that IGF-1 is so rare and expensive makes it highly prized.

The high price of IGF-1 may have so far protected many users from the dangers of overdose. Large doses of IGF-1 will trigger a dramatic drop in blood sugar levels just as surely as an overdose of insulin. And by prolonging the lives of damaged cells that would otherwise die in the body, chronic abuse of IGF-1 could in theory increase the risk of uncontrolled cell growth, and hence cancer. Excess growth hormone has also been linked to side effects, including high blood sugar levels, carpal tunnel syn-

drome and the development of enlarged hands, feet and overly prominent jaws.

Clemmons warns athletes not to use IGF-1 and growth hormone to increase muscle mass because of their "high side effect profile." Typically the dose that is needed to produce anabolic effect on muscles is perilously close to the danger zone. Among the side effects are facial nerve paralysis and a fluid build-up that can cause a dangerous oedema of the brain. "I don't think they'd see a tremendous increase in strength," says Clemmons. "I think they'd start to get themselves into trouble so quickly that they probably would quit doing it or hurt themselves to the point where they couldn't compete."

Athletes intent on abusing anabolic drugs should stick with testosterone, which is cheaper and safer, says Clemmons. "Dollar for dollar, you'd be a whole lot better off taking steroids."

Except, that is, when it comes to drug testing in competitions. For while sports authorities are beginning to test for growth hormone, tests for IGF-1 are still at the planning stages. The International Olympic Committee has banned all peptide hormones in sports, but it doesn't yet have a test for growth hormone, let alone IGF-1. "IGF-1 is certainly on our radar screen, but there are quite a few things on the radar screen," says Donald Catlin, director of the Olympic Analytical Laboratory in the University of California at Los Angeles. Among them is albuterol, an antiasthma medication that has anabolic effects but which clears from the blood stream far more rapidly than steroids.

Growth hormone or IGF-1 are difficult to detect because they are peptides and unlike steroids are not readily excreted into urine. Officials are reluctant to move into blood tests, which would provide a better guide to abuse of growth hormone and IGF-1, because athletes object to this invasive approach. Another issue is deciding what levels of these compounds would indicate abuse. Natural levels of IGF-1 and growth hormone fluctuate throughout the day and vary from one individual to the next.

Officials admit that, given the ever more sophisticated pursuit of performance enhancing drugs, testing, even random testing, has limited effectiveness. But athletes know who the drug users are on their squad or within their sport, says Wade Exum, director of drug control and testing for the US Olympic Committee. "Eventually the athletes will start to expose the cheats." ❏

Questions

1. Why are bodybuilders using the drug insulin-like growth factor-1 (IGF-1)?

2. How are athletes able to get insulin-like growth factor-1 (IGF-1) and growth hormone (GH)?

3. What side effects and other conditions could result from chronic use of IGF-1 and growth hormone (GH)?

Answers are at the back of the book.

43 *Physical activity such as sports or exercise can have beneficial effects on health. In order to maintain the physical gains associated with exercise, your body's need to replace lost fluid becomes crucial to avoid dehydration. While the amount of fluid lost from the body depends on the intensity of the workout and the temperature, the drink you choose to gulp down can also play a role in your overall performance. Drinks that contain carbohydrates and electrolytes such as commercial sports drinks and fruit juices rather than water are the most effective in delaying fatigue. The percent and amount consumed, however, should be geared to the level and duration of physical activity for optimal performance.*

Do Sports Drinks Work?

Trevor Smith

Today's Chemist at Work, June 1996

Drinking water is fine while you're at work, but when you work out and the sweat is rolling, sports drinks are usually the ticket.

"If your muscles are your body's engine, skin is your body's radiator," says exercise physiologist William Evans at Penn State University. Your body's engine produces heat just like your car does. Some of this heat is useful, because it maintains your body temperature constantly at close to 37°C, but you must disperse the rest to avoid overheating.

Your skin can lose heat by radiation, conduction, convection, and evaporation of sweat. There's not much you can do about the first three; they vary with your environment. Evaporation of sweat, however, deserves a closer look. Your skin has sweat glands over most of your body. The most commonly distributed are called eccrine sweat glands; and the larger, higher sweat-producing apocrine sweat glands are found mostly under your arms.

If you are working at your desk with bare arms, they look dry. But if you place the probe of an evaporimeter close to the surface, you find that you are losing water. You are not producing heat rapidly, so the sweat arises from your glands and evaporates from the coils of your sweat ducts to provide a little cooling, and you don't notice the process. If you run upstairs to avoid being late for a meeting, the sudden extra work may make you sweat faster to deal with the extra burst of heat. Evaporation from your sweat ducts can't do the job, so you produce liquid sweat (mostly under your arms, in this case) that collects on the surface to draw more heat out of your body as it evaporates.

There is more to fluid loss than sweat. You excrete fluid as urine, and there is some water in feces. Also, water is one of the oxidation products of carbohydrate and fat, and you lose water continuously while you exhale.

In all, fluid losses are higher than you may think. It is not unusual to wake up as much as a couple of pounds lighter than you were when you went to bed. This is all fluid loss and corresponds to more than thirty ounces, or nearly four cups of fluid. And this is when you are least active.

To keep your physiological and biochemical processes in balance, you must replace fluid losses. How much should you drink? Needs vary with individuals, but the consensus for some years has been that a person should have a minimum of eight glasses of water daily. Many nutritionists today think ten is a better target. Plain water is fine, and so are juices and soft drinks. Although a couple of beers may sound appealing, they help fluid balance less than you think, because alcohol is a diuretic and thus increases your fluid intake and output, too.

Exercise

If you are on a company softball or basketball team, or if you play racket sports at a club, or if you just exercise on your own, your fluid needs become greater. Fluid losses during sports and exercise vary widely among individuals. Aerobic workouts at moderate to hard intensity can make you sweat at a rate of more than thirty ounces an hour. You can check your own sweat rate under various conditions by weighing yourself.

When the weather is very hot, you may sweat so fast that the liquid cannot all evaporate, and excess runs down your skin. If you don't drink to replace the losses, your temperature will rise. If you drink water exactly to balance your losses, your temperature will still rise, but only slightly. The best procedure is to drink as much as you can and try for more intake than what you lose. "Thirst alone is not enough as a guide to how much you should drink," says Evans.

The rate of heat loss is proportional to the temperature difference between your body and its environment. Thus you lose excess heat more slowly in hot weather, unless a faster sweat rate can compensate. But higher humidity in many parts of the country slows down evaporation of sweat. When it is hot and humid, you must drink even more to counteract the effects of faster sweat loss and less effective cooling.

If you fail to compensate for fluid loss, your blood volume decreases, your heart pumps less blood with each beat, and your heart rate increases to try to keep up. Your respiration rate increases, and so does your temperature. These changes soon begin to have a negative effect on your exercise performance. The point at which you notice a problem varies with exercise intensity and rate of fluid loss, but the usual rule of thumb is that about 2 percent of body weight in fluid loss causes significant trouble.

"You may notice a burning sensation in lungs and muscles, shortness of breath, dry mouth, blurred vision, dizziness, and nausea," says athletic trainer Joe Stein, in Lawrenceville, GA. "While you are still capable of rational thought, it is essential to quit exercising, drink plenty of fluid, and get help in case you grow worse. If you don't quit, you may develop dry skin, cessation of sweating, and you will soon approach unconsciousness. At this point your body temperature is out of control and death is not far away." In the early stages, cold drinks and ice massage will correct the problem. Later-stage symptoms call for ice immersion and intravenous fluids which, despite the seriousness of the condition, usually lead to rapid recovery.

Heat exhaustion from dehydration is more common than you may think. Every year, after recreational athletic events such as road races and triathlons, you will see heat victims in the medical tent hooked up to IV bottles. Every year there are vacationers in national parks who need emergency treatment, and some visitors in desert areas such as Big Bend, TX, and the far West have died.

Carbohydrate Solutions

Because the principal component of sweat is water, for many years exercisers were advised that water was the best drink. Experimental evidence said that water was absorbed more quickly than solutions of sugar or carbohydrate. But these studies used sedentary subjects. When physiologists began experiments with people who were exercising, they obtained different results. There are now ample data showing that, during continuous exercise, solutions of carbohydrates up to 6-8 percent concentration are absorbed at the same rate as plain water. Only at higher concentrations does absorption begin to slow. This finding seeded the birth of commercial sports drinks, which are so well established that Gatorade, first in the field, is found in supermarkets.

Do they work? Are there advantages in drinking carbohydrate-containing sports drinks rather than plain water? An overwhelming amount of evidence from many independent laboratories says yes. Consider the examples below.

• When men drank carbohydrate solutions four hours before a cycling test, they performed more work in a forty-five minute workout than they did when they drank water (William Sherman and co-workers at Ohio State University).

• Cyclists given carbohydrate solutions at twelve-minute intervals during a ride performed more work than when they drank water (David Costill and colleagues at Ball State University, Muncie, IN).

• Six cyclists performed a cycling test to fatigue (judged by a specified fall in pedaling rate). Later, they did the same test but were given carbohydrate drinks half an hour before their fatigue point. Five out of six improved their time to fatigue by 22-35 percent (Andrew Coggan and Edward Coyle at the University of Texas at Austin).

• When runners drank a 25 percent glucose polymer solution immediately after a run to exhaustion, their stored glycogen went up rapidly. If they waited two hours before taking the drink, it was only half as effective in replenishing glycogen stores (J. L. Ivy, at the University of Texas at Austin).

• In the 3,000-mile Tour de France race, cyclists consume an average of more than 6,000 calories per day. They take in about half of these while riding, and the principal source is glucose polymer solutions (Wim Saris at the University of Limburg in The Netherlands).

• Larger volumes of carbohydrate solutions leave the stomachs of cyclists faster than small volumes of solution (Joel Mitchell and colleagues at Texas Christian University).

• Men and women began cycle workouts with a 70 percent of maximal effort warm-up. Then they pedaled one-minute bursts at 20-30 percent beyond their anaerobic threshold, followed by three-minute recoveries, and repeated this routine until they could not maintain 70-rpm pedaling rate. When they drank an 18 percent carbohydrate solution before the test and 6 percent carbohydrate solutions every twenty minutes during the test, they stuck with the trial nearly 50 percent longer than when they drank flavored placebos (Russell Pate and coworkers at the University of South Carolina in Columbia).

The mechanism of action of carbohydrate solutions has been studied. About twenty minutes after a person takes a carbohydrate drink during exercise, blood sugar increases. Blood insulin also increases, a sign that the body is ready to metabolize the extra sugar. Then the ratio of carbohydrate to fat oxidized as fuel increases, measured by changes in respiratory exchange ratio (oxygen to carbon dioxide, in and out).

The nature of commercial sports drinks is of little account. They may contain sucrose, glucose, corn syrup, maltodextrin, fructose, and glucose polymers, sometimes singly and sometimes mixtures. The drinks all contain sodium and other minerals that are useful, not because they replace those lost in sweat (the amounts are usually insignificant and are quickly replaced from your food), but because there is convincing evidence that electrolytes promote the absorption of carbohydrate. As far as extending endurance, increasing work output, and restoring glycogen stores are concerned, they are all effective.

For short exercise periods, sports drinks are of little value as far as performance is concerned; you use only a small amount of your carbohydrate stores. However, even for short periods there is an advantage, because studies show that people tend to drink more of flavors they like, and most people prefer sodas and sports drinks to plain water. In other words, you are more likely to avoid dehydration if you drink what you find tastes best.

For optimal performance and to delay fatigue during exercise and sports that last longer than about an hour, you should drink two to four cups of 6-8 percent carbohydrate solutions (either sports drinks or fruit juices diluted fifty-fifty with water) two to four hours before exercise and a cup at twenty-minute intervals during exercise. Immediately after your exercise period, you will benefit from drinks of as much as you can tolerate of 25 percent carbohydrate solutions.

When it is hot and humid, drink all the time; you should never feel thirsty and your urine should be colorless. This is the way to avoid inadvertent dehydration, especially when you work out. ❑

Questions

1. How does your skin lose heat during exercise?

2. Why must you drink more when it is hot and humid?

3. For optimal performance, how much of commercial sports drinks or fruit juices should be consumed?

Answers are at the back of the book.

44 *Watching your weight is important for good health and so is your diet. Studies have been done to demonstrate a connection between lifestyle and risks of breast cancer. Particular foods which offer some protection are olive oil or oils rich in unsaturated fatty acids, fruits, vegetables, fiber and soybeans. Although this information is derived from observational studies, it does help to know we can make enlightened choices about the foods we eat and possibly help to prevent breast cancer.*

Diet and Breast Cancer

Celeste Robb-Nicholson, M.D.

Harvard Women's Health Watch, April 1995

When it comes to breast cancer risk, the news may be getting better. No longer does it seem as though the only factors that influence risk are things we can do little about—family history, having children late or not at all, early menarche, and late menopause; lately, observational studies have begun to demonstrate that diet also may plan an important role.

One of the most recent reports came from Greece, where breast-cancer incidence is lower than in the United States. A team of researchers from the University of Athens and Harvard School of Public Health found that women who consumed olive oil at more than one meal a day had a significantly lower risk of breast cancer than those who used olive oil less frequently. Fruits and vegetables also appeared to have a protective effect.

The study, which appeared in the January 18, 1995, issue of the *Journal of the National Cancer Institute*, with more than 2300 participants, was one of the largest to zero in on specific categories of foods. Women who had breast cancer were matched with women who had similar characteristics but had not developed the disease, and all were asked to report how often they ate several types of foods, including margarine, olive oil, and animal fats.

As was the case in many other studies looking for links between lifestyle and cancer, the information depended upon the women's ability to accurately report the foods they ate. Because memory is fallible, such observational studies are considered to be less definitive than controlled trials in which two groups of women with similar characteristics are placed on different diets and the number of new breast cancers in each group is compared. Nonetheless, observational studies are valuable because they not only focus the direction of future clinical trials but also serve as interim guidance for those of us trying to do what we can to prevent breast cancer.

To date, observational studies have indicated certain connections between breast cancer and the following foods:

• *Dietary fat.* The evidence is conflicting. Although some studies show breast cancer risk increasing with fat intake, the overall breast cancer rate is significantly lower among women in the Arctic North and the Mediterranean, where fat accounts for more than 40% of calories, than among American women, who consume, on average, 36% of calories as fat.

In both of the latter populations, fat is provided primarily by fish oils or olive oil, both of which are rich in unsaturated fatty acids. In contrast, dietary fat for most Americans is derived principally from saturated fatty acids in meat and dairy products and from hydrogenated polyunsaturated oils containing *trans*-fatty acids, which are found in margarine and packaged snacks and pastries. Because there is no indication that a low-fat diet increased breast cancer risk, a prudent course would be to restrict fat to 30% of calories, as recommended by the National Research Council, by reducing saturated fats and *trans*-fatty acids.

• *Dietary fiber.* A high fiber intake is consistently associated with a lowered risk of breast cancer. Insoluble fibers like bran, derived principally from unprocessed grains, seem to provide the greatest protective effect; soluble fibers from fruits and vegetables also seem to confer protection.

Fiber intake appears to reduce levels of circulating estrogens, which may promote the growth of breast tumor, perhaps by curtailing the production of liver enzymes that foster estrogen absorption and by binding to the estrogens in the intestines.

• *Soybeans*, which contain cancer-preventing compounds called isoflavonoids, are also a source of phytoestrogen, which is chemically similar to estrogen. By replacing estrogen on certain tissue receptors, phytoestrogens may block its cancer-promoting effects. ❏

Questions

1. What level of fat is acceptable in the diet according to the National Research Council?

2. What health benefit does fiber provide?

3. What cancer-preventing compound is found in soybeans?

Answers are at the back of the book.

Fat substitutes in food could be a godsend to the traditional American diet of high fat and energy content. While food chemists strive to develop a substance which behaves like fat without the calories or cholesterol, the public craves to know if fat replacement is a good idea. It may be too early to tell if the use of fat-reduced foods will change the amounts and types of food consumers will choose and whether their choices will produce the desired effects of reducing fat intake. As the list of current and proposed ingredients for replacement grows, the information from dietary intervention trials of the effects of using reduced-fat food should answer some fundamental questions about fat replacement in the diet.

Implications of Fat Replacement for Food Choice and Energy Balance

David J. Mela

Chemistry & Industry, May 6, 1996

It seems like the perfect dietary solution: fat substitutes that don't make you fat. But what effect will they really have on fat intake and weight control?

The recent US Food & Drug Administration approval of olestra for limited use as a fat substitute[1] gave a high profile to the divergent public views on the application of technology in foods. For some individuals and groups, olestra is seen as another unwanted and potentially dangerous technological fix for a faulty food supply system and poor individual eating habits. For many others, and for many manufacturers, it is the Holy Grail of food ingredients: a substance with the feel and function of fats, but without the calories or cholesterol. Public concern about the role of fat in obesity and chronic diseases has stimulated a massive flow of resources into the development and marketing of alternative processing techniques and nutritive or nonnutritive materials which can be used to remove or replace fat in foods.

While food scientists have largely focused on the functionality of existing and proposed ingredients and processes for fat replacement, and the regulators have scrutinised the replacements' safety, the overall concept has been sold on the implicit premise that fat replacement in foods will deliver an important nutritional benefit. But it is probably fair to say that we have much more established knowledge of the shelf life, rheological behaviour, and toxicology of specific fat replacement technologies, than about their likely effectiveness for reducing fat intake or for aiding in weight control. However, recent advances in our understanding of the role of fat in appetite and eating behaviour, and a number of experimental and consumer trials of reduced-fat foods and diets, have begun to provide the information needed to address these issues.

Ingredients for Fat Replacement

Some examples of current and proposed ingredients for fat replacement are listed in the Table; their functional and sensory properties and limitations are discussed in detail in other sources.[2-3] The physiological effects of these different materials raise some particular nutritional issue,[4] and their functionality will have a significant bearing on the range and extent of fat replacement which might occur across the diet.

Materials based on carbohydrates and proteins are biologically well understood, and are used to provide fat-like textures in a wide range of products, although they have very significant limitations as general fat replacers. These materials create a reduced calorie product by virtue of their low energy density relative to fats, particularly when fully hydrated in food products. Some, such as cellulose and certain gums, are truly resistant to digestion and can act as bulking agents within the gut. These properties of non-starch polysaccharides are well known and are not unique to their use as fat replacers. In fact, they may have an independent influence on appetite.[5]

Replacements based on lipids, the other major category, have functional and sensory properties very similar to the fats they might replace, including the characteristic texture and flavour effects of native fats. Several of these materials are completely resistant to hydrolysis in the gut and are excreted intact, thereby effectively contributing no usable calories or fat to the diet. Although they could work satisfactorily in almost all current cooking oil applications, concerns about their physiological and nutritional effects, particularly at high intake levels, have been an obstacle to the approval and food use of such completely non-absorbably lipids.[2-4] Only one of these, a mix of hexa-, hepta- and octa- fatty acid esters of sucrose ('sucrose polyester', now assigned the general name olestra), has reached the stage where limited food use (in savoury snacks) has been approved in the US.[1]

Several more recently developed materials take advantage of the limited absorption of naturally-occurring long-chain saturated fatty acids, and the lower caloric density of medium and short chain fatty acids, generating structured triglycerides with reduced caloric value.[6,7] These would typically have a metabolisable energy value of about 4-5kcal/gm, compared with about 9kcal/gm for ordinary mixed dietary fats. At least one of these types of products, caprenin, is currently used in selected food applications, particularly confectionery.[3]

Polyglycerol esters and a range of other emulsifiers and stabilisers form an intermediate category. Depending on the material, their functional and metabolic properties may vary quite widely. These ingredients can be partially or wholly digestible, but are generally used (as part of an aqueous emulsion or suspension) at rather low concentrations relative to the quantity of fat they might replace.[8]

Role of Fat in Satiety, Overeating and Obesity

A growing body of evidence from metabolic, experimental and epidemiological studies suggests that a high relative fat intake might specifically contribute to overeating and obesity, more so than other dietary energy sources. Support for this notion comes from a number of observations.

Fat is energy dense, but may have a weak effect on satiety. At 9kcal/gm, fat is considerably more energy dense than protein and carbohydrates (both around 4kcal/gm), although it is important to point out that this fact alone does not implicate fat in overeating and obesity. But sedentary humans and other animals seem less able to control their energy balance in the face of high energy density.[9] Furthermore, studies by John Blundell's group at the University of Leeds, and others, provide support for the view that calories from fat can exert a disproportionately weak ability to suppress eating during the meal or at subsequent meals, at least in certain subgroups.[10-13] That is, individuals tend to eat more of high fat versus low fat meals, even when equated for palatability, and this excess intake may not be fully compensated for in subsequent meals. Other studies also confirm that humans may be poor at adjusting intakes in response to covert increases in food energy largely made up of fat.[14,15]

Increased intakes of fat do not stimulate fat oxidation. An extensive range of animal and human metabolic and feeding trials indicate that while carbohydrate oxidation is closely linked to carbohydrate intake, fat oxidation is largely unrelated to fat intake.[16-18] Dietary fat is therefore largely stored and used later on to make up the difference between energy expenditure and energy generated from other sources. Further experiments have confirmed and extended these findings, and it appears that a predisposition to obesity may be associated with defects in the ability to raise fat oxidation commensurate with moderate or high fat intakes.[19,20] Some researchers, including Arne Astrup and colleagues at the Royal Veterinary and Agricultural University, Denmark, have proposed that obesity in susceptible individuals is an adaptive response to a high fat diet, with the necessary equilibration of fat intake and oxidation being achieved by expanding body fat stores.[20,21]

Dietary fat is very efficiently converted to body fat. Laboratory studies, feeding trials and dietary intervention studies corroborate the metabolic evidence, and show that dietary fats are more efficiently used and more readily converted to body fat than carbohydrate, the other main energy source in the diet, perhaps even at levels of energy intake intended to achieve comparable body weights.[22-24] Furthermore, in clear contrast to traditional views of human energy metabolism, there is now doubt as to whether conversion of dietary carbohydrates to body fat (*de novo* lipogenesis) actually occurs to any significant extent under normal feeding conditions in humans.[25-28]

Higher fat intakes are associated with greater relative body weight in population studies. Although the relationship is complex, a number of epidemiological studies and dietary sur-

veys find positive associations between the amount of dietary energy derived from fat and measures of overweight and obesity.[29] Longitudinal data also point to causal relationships between fat intakes and weight gain over time, at least in susceptible individuals.[30,31]

Subjects consuming low fat diets *ad lib* tend to undereat and lose weight. Several experimental trials have found that an extended shift to a strict low fat (but not calorie restricted) diet is typically accompanied by spontaneous reductions in energy intakes or weight loss, even when these are not part of the specific dietary guidance.[23,32-34] Examination of many of the larger clinical trials of reduced-fat diets also reveals weight loss as a side-effect of regimens intended to improve blood lipid profiles. Conversely, Lauren Lissner and colleagues showed in studies at Cornell University, New York, that consumption of high fat diets under relatively free-living conditions could produce overeating and weight gain.[15]

Implications for Fat/Energy Intake

It would seem, then, that replacing fat in foods should help prevent and treat obesity. However, this depends on the amount of fat replaced, and the general response to the sustained consumption of reduced fat and energy versions of common foods. Data relevant to this issue come from a number of relatively small-scale studies in which the energy density of foods has been (usually covertly) manipulated by substitutions for fat and/or sugar.[35,36] Most of this research has been conducted within the confines of a laboratory or clinic, or on free-living subjects provided with fixed meals or limited food selections, and generally over no more than a 1-14 day period.

Studies to date have largely shown that, except where the manipulation is very great or food selection is restricted, humans generally compensate fairly well for dilutions of food energy. However, this is not always seen.[37] Most or all of the 'lost' energy is compensated for by increased consumption of the same or other foods—so it seems that compensation isn't macronutrient specific (but see reference 13). However, few studies have specifically addressed these issues with subjects in a more normal setting over an extended period; furthermore, compensation for reduced energy foods could be further enhanced by knowledge of the manipulation behind it (for example, influences from conventional labels, packaging and advertising claims). Information on the nutrient composition of foods has been found to have important effects in certain experimental situations,[38,39] but very little in others.[37] The specific nature of the information provided and its influences on the behaviour of consumers with different characteristics are potentially important issues for further research.

At the Institute of Food Research, we have carried out dietary intervention trials to examine the effects of using reduced-fat food within the natural settings for food purchase and consumption.[40,41] Free-living consumers bought their own foods in retail supermarkets, and ate them *ad lib* at home over six

---

Selected Examples of Current and Proposed Ingredients for Fat Replacement

Carbohydrate and Protein-based Materials
Modified glucose polymers
Modified corn, potato, oat, tapioca and rice starches
Gums and algins
Cellulose and cellulose derivatives
Gelatin
Microparticulated proteins

Poorly Absorbed or Non-absorbable Lipids
Fatty acid esters of sugars and sugar alcohols (for
    example, olestra)
Structured lipids containing specific fatty acids (for
    example, caprenin)
Polycarboxylic acid and propoxylated glyceryl esters
Alkyl glyceryl ethers
Substituted siloxane polymers
Branched (sterically hindered) triglyceride esters
Specific naturally occurring lipids

Emulsifiers and Functional Ingredients
For example, polyglycerol esters, lecithins and milk proteins

---

or ten weeks. Relative to controls, subjects who were advised to make extensive substitutions of reduced-fat for full-fat foods attained a substantial reduction in the percentage of energy from fat. These normal-weight subjects spontaneously increased their intakes of other macronutrients, so that overall energy intakes did not differ significantly from those of the controls.

Consumer Behaviour Issues
Food choices/dietary patterns. Little is known about the influence that existing or future fat replacement might have on overall food selection, and the issue warrants attention as reduced-fat products proliferate in the market. Concern has been expressed that increased interest could be directed towards a class of foods that some consumers currently avoid because of the high fat and energy content. Consumption of these products could therefore increase at the expense of other, more nutrient-dense items.[42,43] Could reliance upon modified foods undermine efforts to promote a wider range of desirable dietary and nutritional goals? This scenario remains speculative as there are few if any relevant data addressing the issue. Particular attention may need to be paid to groups with special needs (for example, children and older people), or who might be expected to incorporated fat-substituted foods into their diets at relatively high levels or are predisposed to eating disorders (such as adolescent females).

Effectiveness in use. Could consumers respond to the use of reduced-fat foods by consciously or unconsciously increasing their selection of fat from other sources? UK National Food Survey data suggest that reductions in the consumption of specific fat sources (for example, butter, lard and whole milk) have had little effect on the relative fat intake of the population, as these changes are offset by increased consumption of fat derived from other foods.[44] Our own data suggest that many consumers may also have unrealistic views about the extent to which they have already reduced their fat intakes, believing that they have already made reductions to the point where further changes may not be necessary.[45,46] The extent to which the use of fat-reduced foods might contribute to such misconceptions, and perhaps resistance to complementary dietary changes, is not known.

Florence Caputo and Rick Mattes carried out work at the Monell Chemical Senses Center in Philadelphia, showing that subjects increased their total freely-selected fat intakes during test periods in which they were provided with what they believed to be lower-fat lunch meals.[38] Under real eating conditions, such cognitive effects could counteract the beneficial effects revealed under rigorously controlled laboratory conditions. However, we found that the use of reduced-fat foods substantially lowered fat intake among consumers purchasing these items with full knowledge and commercial information.[40]

Food volume and costs. One consumer issue which is rarely addressed is the potential cost associated with the versions of foods with reduced fat and energy. Assuming that these specialised products cost as much as or more than their traditional counterparts, and that their use is accompanied by at least partial caloric compensation, it necessarily follows that the total volume of food consumed, and hence the cost to the consumer, must increase to sustain this level of caloric intake. While this could be seen as an economic benefit to the food industry, it may represent a consumer issue of the future.

Long-term acceptance. It also cannot be assumed that versions of foods severely reduced in fat and energy will maintain their acceptability to consumers even if these products retain the sensory characteristics of their full-fat counterparts. Preferences for dietary fats in general or for particular high-fat food items may be largely linked to effects of fats or fat metabolites in the gut or on subsequent metabolic processes.[47] That is, some aspect of the metabolism of fats may be a critical component of a learned preference process. Hence, 'palatability' and food preferences cannot be considered to be unrelated to the nutritional and compositional aspects of foods. This raises fundamental questions about the ultimate, long-term consumer acceptance of very

low calorie, fat-replaced foods, which disassociate the sensory and metabolic properties of fats.

Conclusions

Fat replacement ostensibly offers consumers the possibility of enjoying tasty and nutritional versions of many popular foods. The use of reduced-fat foods could therefore offer opportunities for altering intakes within the existing traditional diet, and act as a potentially useful and acceptable adjunct to broader strategies for achieving changes in the types and composition of foods consumed. However, at the moment there is only limited information with which to assess the likely impact of these products on food intake and overall diet quality. Based on the available literature, a number of tentative conclusions can be suggested.

Widespread incorporation of fat replacers into foods could make a significant contribution to reducing total and relative fat intake, although the overall magnitude is not clear. Fat intake appears to be clearly related to overeating and obesity, so substantial reductions in fat content and energy density of the diet would seem likely to help control energy intake and body weight. However, large-scale fat replacement in food products may have unanticipated effects on food selection and acceptance, issues which have not yet been examined in any detail.

The effects of fat replacement in foods will ultimately reflect the overall behaviour of consumers, the shifts in food choice, and the actual patterns of nutrient and energy intakes resulting from informed purchase and consumption of commercial products.

Acknowledgements
Support from the UK Biotechnology & Biological Sciences Research Council is gratefully acknowledged.

References
1. Lindley, M., *Food Manufacture*, 1996, **71(3)**, 51-53, 60
2. Iyengar, R., & Gross, A., in 'Biotechnology and food ingredients' (Eds I. Goldberg & R. Williams), *New York: van Nostrand Reinhold*, 1991,

287-313

3. Lindley, M.G., in 'Low-calorie foods and food ingredients' (Ed. R. Khan), *London: Blackie*, 1993, 77-105

4. Mela, D.J., *Fett/Lipid*, 1996, **98**, 50-55

5. Burley, V.J., & Blundell, J.E., in 'Dietary fiber in health & disease' (Eds D. Kitchevsky & C. Bonfield), *St. Paul, MN: Eagan Press*, 1995, 243-56

6. Finley, J.W., Klemann, L.P., Levielle, G.A., Otterburn, M.S., & Walchak, C.G., *J. Agric. Food Chem.*, 1994, **42**, 474-83

7. Peters, J.C., Holcombe, B.N., Hiller, L.K., & Webb, D.R., *J. Am. Coll. Toxicol.*, 1991, **10**, 357-67

8. Lucca, P.A., & Tepper, B.J., *Trends Food Sci. Technol.*, 1994, **5**, 12-19

9. Poppitt, S.D., *Int. J. Obesity*, 1995, **19(Suppl 5)**, S20-S26

10. Blundell, J., Burley, V., Cotton, J., & Lawton, C., *Am. J. Clin. Nutr.*, 1993, **57(suppl)**, 772S-778S

11. Blundell, J.E., *et al*, *Int. J. Obesity*, 1995, **19**, 832-5

12. Lawton, C.L., Burley, V.J., Wales, J.K., & Blundell, J.E., *Int. J. Obesity*, 1993, **17**, 409-16

13. Rolls, B.J., *et al*, *Am. J. Clin. Nutr.*, 1994, **60**, 476-87

14. Caputo, F.A., & Mattes, R.D., *ibid*, 1992, **56**, 36-43

15. Lissner, L., Levitsky, D.A., Strupp, B.J., Kalkwarf, H.J., & Roe, D.A., *ibid*, 1987, **46**, 886-92

16. Flatt, J.P., *Ann. NY Acad. Sci.*, 1993, **683**, 122-40

17. Swinburn, B., & Ravussin, E., *Am. J. Clin. Nutr.*, 1993, **57(suppl)**, 766S-771S

18. Schutz, Y., Flatt, J.P., & Jéquier, E., *ibid*, 1989, **50**, 307-14

19. Astrup, A., *Int. J. Obesity*, 1993, **17(Suppl 3)**, S32-S36

20. Schutz, Y., *Obesity Res.*, 1995, **3(Suppl 2)**, 173s-178s

21. Astrup, A., *et al*, *Am. J. Clin. Nutr.*, 1994, **59**, 350-5

22. Danforth, E., *ibid*, 1985, **41**, 1132-45

23. Sheppard, L., Kristal, A.R., & Kushi, L.H., *ibid*, 1991, **54**, 821-8

24. Prewitt, T.E., *et al*, *ibid*, 1991, **54**, 304-10

25. Acheson, K.J., *et al*, *ibid*, 1988, **48**, 240-7

26. Björntorp, P., & Sjöström, L., *Metabolism*, 1978, **27**, 1853-65

27. Hellerstein, M.K., *et al*. *J. Clin. Invest.*, 1991, **87**, 1841-52

28. Weiss, L., *et al*. *Biol. Chem. Hoppe-Seyler*, 1986, **367**, 905-12

29. Lissner, L., & Heitmann, B.L., *Eur. J. Clin. Nutr.*, 1995, **49**, 79-90

30. Klesges, R.C., Klesges, L.M., Haddock, H.K., & Eck, L.H., *Am. J. Clin. Nutr.*, 1992, **55**, 818-22

31. Heitman, B.L., Lissner, L., Sørensen, T.I.A., & Bengtsson, C., *ibid*, 1995, **61**, 1213-17

32. Jeffery, R.W., Hellerstedt, W.L., French, S.A., & Baxter, J.E., *Int. J. Obesity*, 1995, **19**, 23-137

33. Kendall, A., Levitsky, D.A., Strupp, B.J., & Lissner, L., *Am. J. Clin. Nutr.*, 1991, **53**, 1124-9

34. Shah, M., McGovern, P., French, S., & Baxter, J., *ibid*, 1994, **59**, 980-4

35. Bellisle, F., & Perez, C., *Neurosci. Biobehav. Rev.*, 1994, **18**, 97-205

36. Mela, D.J., in 'Progress in obesity research 7' (Eds. A. Angel, H. Anderson, C. Bouchard, D. Lau, L. Leiter, & R. Mendelson), *London: John Libbey & Co.*, 1966, 423-30

37. Hulshof, T., 'Fat and non-absorbable fat and the regulation of food intake', PhD thesis, Wageningen Agricultural University, The Netherlands, 1994

38. Caputo, F.A., & Mattes, R.D., *Int. J. Obesity*, 1993, **17**, 241-4

39. Shide, D.J., & Rolls, B.J., *J. Am. Diet. Assoc.*, 1995, **95**, 993-8

40. Gatenby, S.J., Aaron, J.I., Morton, G., & Mela, D.J., *Appetite*, 1995, **25**, 241-52

41. Aaron, J.I., Gatenby, S.J., Jack, V., & Mela, D.J., *Proc. Nutr. Soc.*, in press (abstract)

42. Munro, I.C., *Food Chem. Toxicol.*, 1990, **28**, 751-3

43. Owen, A.L., *J. Am. Diet. Assoc.*, 1990, **90**, 1217-22

44. Ministry of Agriculture, Fisheries and Food,

'National food survey 1992', *London: HMSO*, 1993

45. Lloyd, H.M., Paisley, C.M., & Mela, D.J., *Eur. J. Clin. Nutr.*, 1993, **47**, 361-73

46. Paisley, C.M., 'Barriers to the adoption and maintenance of reduced-fat diets', PhD thesis, University of Reading, UK, 1994

47. Mela, D.J., *Proc. Nutr. Soc.*, 1995, **54**, 453-64 ❏

Questions

1. What is harmful about dietary fat and fat consumption?

2. How does dietary fat compare metabolically with other energy components in the diet?

3. What implications will fat replacement have on fat/energy intake?

Answers are at the back of the book.

Section Seven

Controversies and Ethical Issues

46 *It may sound like science fiction, but xenotransplants are the organ transplants of the future. For patients with severe kidney failure, liver disease, heart defects and other diseases, an organ from an animal could be their only hope for survival. The need to use animals as donors stems from the fact that there are simply not enough human organs to go around. For this reason, surgeons are perfecting their transplant techniques to prepare for the organ shortage. Although still in the experimental stages, xenotransplantation must clear several crucial hurdles, one of which is the possibility of an outbreak of a new infectious disease. This and other serious risks loom on the horizon as scientists explore this new frontier.*

Xenotransplants from Animals: Examining the Possibilities

Rebecca D. Williams

FDA Consumer, June 1996

"You'll need a liver transplant," Dr. Zeno says. She scribbles quickly on her prescription pad and dates it: April 17, 2025. "Take this to the hospital pharmacy and we'll schedule the surgery for Friday morning."

The patient sighs—he's visibly relieved that his body will be rid of hepatitis forever.

"What kind of liver will it be?" he asks.

"Well, it's from a pig." Zeno replies. "But it will be genetically altered with your DNA. Your body won't even know the difference."

Obviously, this is science fiction. But according to some scientists, it could be a reality someday. An animal organ, probably from a pig, could be genetically altered with human genes to trick a patient's immune system into accepting it as its own flesh and blood.

Called "xenotransplants," such animal-to-human procedures would be lifesaving for the thousands of people waiting for organ donations. There have been about 30 experimental xenotransplants since the turn of the century.

Rebuilding Bodies

Xenotransplants are on the cutting edge of medical science, and some scientists think they hold the key not only to replacing organs, but to curing other deadly diseases as well.

Last December, for example, after getting permission from the Food and Drug Administration, researchers at the University of California, San Francisco, injected an AIDS patient with baboon marrow. The hope was that the baboon marrow, which is resistant to HIV and a source of immune cells, could provide a replacement for the patient's damaged immune system.

In April 1995, also with FDA permission, doctors at Lahey Hitchcock Medical Center in Burlington, Mass., injected fetal pig brain cells into the brains of

Reprinted from *FDA Consumer*, June 14, 1996.

patients with advanced Parkinson's disease. The hope was that the fetal tissue would produce dopamine, which the patients' brains lack. Both experiments were primarily to test the safety of such procedures, not whether they are effective.

Other xenotransplant experiments have involved implanting animal hearts, livers and kidneys into humans.

According to Scott McCartney's book on transplantation, *Defying the Gods: Inside the New Frontiers of Organ Transplants*, the first organ transplant was performed in the early twentieth century by Alexis Carrel, a French physician practicing in Chicago. He had developed a technique to sew blood vessels together, and in 1906 he transplanted a new heart into a dog and a new kidney into a cat.

The first animal-to-human transplant was in the same year, when the French surgeon Mathieu Jaboulay implanted a pig's kidney into one woman and a goat's liver into another. Neither survived.

Today, human organ transplants are commonplace. For example, more than 10,000 Americans received kidney transplants last year, with a three-year life expectancy of more than 85 percent, according to the United Network for Organ Sharing (UNOS), an organization of transplant programs and laboratories in the United States. Under contract to the U.S. Department of Health and Human Services, UNOS administers a national organ network, and its members set policies for equitable organ allocation.

Surgeons have made great strides in perfecting transplant techniques, but two problems endure. First, there are never enough organs to go around. Second, once patients receive organs, it is a constant battle to keep their immune system from rejecting them. Both problems may be eventually solved by xenotransplants and the genetic engineering techniques developed from such experiments.

Of all animals, baboons and pigs are the favored xenotransplant donors. Baboons are genetically close to human, so they're most often used for initial experiments. Six baboon kidneys were transplanted into humans in 1964, a baboon heart into a baby in 1984, and two baboon livers into patients in 1992.

Although all the patients died within weeks after their operations, they did not die of organ rejection. Rather, they died of infections common to patients on immunosuppressive drugs.

One drawback to using baboons is that they harbor many viruses. They also reproduce slowly, carrying only one offspring at a time. Some people have raised ethical objections, especially since baboons are so similar to humans. They have human-like faces and hands and a highly developed social structure. Although it's conceivable that baboons could donate bone marrow without being killed, recent experiments have required extensive tissue studies, and the animals have been sacrificed.

For long-term use, pigs may be a better choice. Pigs have anatomies strikingly similar to that of humans. Pigs are generally healthier than most primates and they're extremely easy to breed, producing a whole litter of piglets at a time. Moral objections to killing pigs are fewer since they're slaughtered for food.

Pig organs have been transplanted to humans several times in the last few years. In 1992, two women received pig liver transplants as "bridges" to hold them over until human transplants were found. In one patient, the liver was kept outside the body in a plastic bag and hooked up to her main liver arteries. She survived long enough to receive a human liver. In the other patient, the pig liver was implanted alongside the old diseased liver, to spare the patient the rigors of removing it. Although that patient died before a human transplant could be found, there was some evidence that the pig liver had functioned for her.

By genetically altering pig livers, some scientists believe they can make a pig liver bridge more successful. In July 1995, FDA permitted the Duke University Medical Center to test genetically altered pig livers in a small number of patients with end-stage liver disease. The pig livers contained three human genes that will produce human proteins to counter the rejection process.

Safe or Disastrous?

Xenotransplantation could be very good news for patients with end-stage organ diseases. There would be no more anxious months of waiting for an organ donor. Disease-free pigs would provide most of the organs. Raised in sterile environments, they would be genetically altered with human DNA so that the chance of rejection is greatly reduced.

Transplant surgery would be scheduled at the patient's convenience, as opposed to emergency surgery performed whenever a human donor is found. Patients wouldn't have to wait until their disease were at a critical stage, so they would be stronger for recovery.

Today, however, xenotransplantation is still experimental, and there are serious risks to the procedures.

Although many researchers believe it is slight, one legitimate concern is that animal diseases will be transmitted into the human population. Baboons and swine both carry myriad transmittable agents that we know about—and perhaps many more we cannot yet detect. These bacteria, viruses and fungi may be fairly harmless in their natural host, a baboon or pig, yet extremely toxic—even deadly—in humans.

The two types of animal viruses that are especially troublesome are herpes viruses and retroviruses. Both types have already been proven to be rather harmless in monkeys, but fatal to humans. HIV, for example , is a retrovirus that many researchers believe was transmitted to humans from monkeys. The problem occurs in reverse as well. Measles, for example, a serious but manageable disease in humans, can destroy a whole colony of monkeys quickly.

By regulating xenotransplants, FDA will provide a framework of collecting safety data and tracking patients' health. The process should involve open and public discussion by scientists about their experiments, allowing their peers to evaluate and critique them, and their patients to understand the risks and make informed decisions.

"Will [xenotransplants] cause an outbreak of a new infectious disease? We don't know," says Phil Nogouchi, M.D., a pathologist and director of FDA's division of cellular and gene therapies. "But we want all these procedures discussed in public. We need to make people aware of the hazards."

Nogouchi emphasizes the importance of monitoring and tracking all recipients of xenotransplants so that if any new diseases do develop, they will be detected quickly and the threat to public health will be minimized.

"We cannot say that's not a possibility," says Nogouchi. "But we do feel the potential benefits are great and that efforts can be made to make everyone responsible. There are ways to deal with problems should they arise."

At press time, FDA, the National Centers for Disease Control and Prevention, and the National Institutes of Health were working on recommendations for researchers doing xenotransplant experiments.

Although the new recommendations will be for researchers, patients will likely also recognize their importance.

"Our biggest allies are the patients," says Nogouchi. "They should be asking, 'Where'd you get that pig?'" Xenotransplants cannot be "fresh off the farm." They should be bred and raised in a biomedical animal facility under strict conditions.

Battling Rejection

The other formidable obstacle to xenotransplants is that posed by the human body's own immune system. Even before a person is born, his or her immune system learns to detect and resist foreign substances in the body called antigens. These could be from anything that's not supposed to be there: viruses, bacteria, bacterial toxins, any animal organs, or even artificial parts.

Antigens trigger the body's white blood cells, called lymphocytes, to produce antibodies. Different lymphocytes recognize and produce antibodies against particular antigens. B cell lymphocytes produce antibodies in the blood that remove antigens by causing them to clump or by making them more susceptible to other immune cells. T cell lymphocytes activate other cells that cause direct destruction of antigens or assist the B cells.

Transplant physicians try to suppress the immune system with powerful drugs. While these drugs are often successful, they leave the patient vulnerable to many infections. FDA-approved immunosuppressive drugs include Sandimmune (cyclosporine), Imuran (azathioprine), Atgam (lymphocyte immune globulin), Prograf (tarolimus), and Orthoclone (muromonab-CD3). New drugs are also being researched, including some "designer" immune suppressants. These drugs may enable doctors to suppress the immune system from rejecting a particular organ, but leave the rest of the body's immune system intact.

Drugs designed to help transplant patients may end up also aiding those who are stricken with diseases such as arthritis, multiple sclerosis and diabetes, because these involve problems with the human immune system. For example, Imuran is approved to treat severe rheumatoid arthritis, and Prograf has already shown some promise to MS patients. A large study is under way to determine if it is effective.

Genetic engineering is the next step in battling organ rejection. Researchers have begun experimenting with ways to insert human genes into animal organs, so that the organs will produce proteins the body will recognize as "human." FDA is active in basic research that may lead to better gene therapies and ways of manipulating animal organs.

For example, Judy Kassis, Ph.D., an FDA biochemist, has been studying a fruit fly gene that is important to the insects' early development. Using some DNA and a harmless virus, she has developed a way to insert this gene precisely into its natural position on the fly's chromosomes. Carolyn Wilson, Ph.D., and FDA virologist, has been researching pig viruses and whether they could infect humans in a transplant setting.

FDA scientists are also studying ways that individual genes "turn on" as they develop, how viruses activate each other, and how viruses can be used safely to deliver genes for new therapies.

"Gene therapy is really in its infancy," says Kassis. "That's the thing about basic research—you can't really predict how useful this will be in the future. Hopefully, it will have direct relevance someday."

Gene therapies and their role in xenotransplantations are still in the early stages of development. For now, it's only in science fiction that doctors can order a custom-designed pig liver from the hospital pharmacy. Whether or not that ever becomes reality, FDA's goal in regulating xenotransplant experiments is to make sure these procedures are openly discussed, that data is carefully collected, that patients give their fully informed consent, and that safety precautions are taken with every effort. ❑

Questions

1. What are xenotransplants?

2. Which animals are most favored for xenotransplantation?

3. What are two drawbacks to animal organ transplants?

Answers are at the back of the book.

47

Even with increased public awareness among women of the benefits of hormone replacement therapy, controversy stills exists about its safety. Unequivocal benefits to the use of estrogen in controlling menopausal systems while decreasing the risk of heart disease and osteoporosis are well documented. The fear that estrogen replacement may be linked to the development of uterine and breast cancer, however, is an issue of grave concern to the average American woman and the subject of massive clinical research. Until the results of the study are in, women will need to weight the potential benefits of hormone therapy on heart disease and osteoporosis against a possible increased risk of cancer.

Is Hormone Replacement Therapy a Risk?

Nancy E. Davidson

Scientific American, **September 1996**

It is largely fear of breast cancer that fuels the debate about hormone replacement.

Thanks to advances in public health and medicine, the average American woman will be postmenopausal for about one third of her life. As a result, she will ultimately need to make a decision about hormone replacement therapy. During the 1960's, doctors began to prescribe a short-term regimen of estrogen to control menopausal symptoms such as hot flashes and vaginal dryness. More recently, physicians have realized that long-term use can reduce illness and death from heart disease and bone loss (osteoporosis). These potential benefits, however, are balanced to some extent by a possible increased risk of cancer, especially of the breast and uterus.

Indeed, it is largely fear of breast cancer, the most common cancer in women in the U.S., that fuels the debate about hormone replacement. But in weighing the risks and benefits, we must recall that heart disease is the most prevalent cause of death for American women. In 1992 approximately 250,000 women died of coronary disease. Cancer ran a close second at 245,000 deaths for all types; the top three—lung, breast and colorectal cancer—account for 55,000, 43,000 and 29,000 deaths, respectively.

How much do we know about the impact of hormone replacement therapy on heart disease, osteoporosis and cancer? A number of studies have suggested that the use of estrogen for several years decreases risk of heart disease by up to 50 percent—a critical finding in view of the prevalence of coronary disease among women in this country. Long-term hormone therapy also appears to be valuable in preventing the bone fractures that stem from osteoporosis. Hip fractures, which afflict over 175,000 women in the U.S. every year, can destroy vitality, lower the quality of life and lead to death. Sustained use of estrogen appears to reduce hip fractures by 30 to 40 percent; fractures at other sites seem to decrease as well. Furthermore, preliminary evidence hints that the therapy may offer some degree of protection against Alzheimer's disease. Its usefulness in preserving function of the genitourinary tract and in preventing tissue atrophy is well documented.

Most research shows that the greatest benefits of estrogen replacement come with continuous use that begins shortly after menopause. The bone-protecting effects, in particular, diminish rapidly within a few years of stopping medications. Unfortunately, this need for long-term use raises the fear that estrogen replacement might also be linked to the development of two hormonally related cancers, uterine and breast cancer.

Unequivocal evidence suggests that estrogen therapy increases the risk of uterine cancer by up to sixfold over that seen in women who do not take estrogen. Uterine cancer, however, is usually diagnosed early, and thus many deaths from the disease can be prevented (about 6,000 women die for this type of cancer every year). Even more important, the addition of another hormone, a progestin, markedly lessens the possibility of uterine cancer. This finding has led to the frequent prescription of estrogen and progestin together as a means of trying to maintain the cardiac and bone benefits of estrogen without increasing the likelihood of uterine cancer.

What about the effects of hormone replacement on breast cancer? That breast cancer is in part hormonally mediated is known from extensive epidemiologic studies. But the connection between breast cancer and hormonal therapy is not clear. Several dozen studies of various types have yielded mixed results. In aggregate, they suggest that less than five years of estrogen therapy has no impact on breast cancer. Some studies, however, show that the risk of breast cancer increases by 15 to 40 percent after longer durations of estrogen replacement, with or without progestin. Thus, long-term replacement, which has optimal effects on heart disease and osteoporosis, may well be linked to a small increase in the incidence of breast cancer.

A little known finding is that hormone replacement therapy appears to offer some protection against another deadly malignancy, colon cancer. Several studies now indicate that women taking hormone replacement therapy have half the chance of dying from colon cancer when compared with those who are not taking hormones.

Given the uncertainty about the exact impact of this therapy, the National Institutes of Health has launched a 15-year, nationwide clinical trial involving postmenopausal women. Called the Women's Health Initiative, it will evaluate the total health effects of hormone replacement therapy. Women who have had a hysterectomy and therefore have no risk of uterine cancer will be randomly assigned to daily estrogen or a placebo; those with an intact uterus will be assigned to daily estrogen plus progestin or will be given a placebo. This trial will focus on heart disease and osteoporotic fractures, but information about breast and colon cancer may also emerge, with the earliest findings expected at the beginning of the next century. [For information on how to participate, call (800) 549-6636.]

In the meantime, women must be guided by their own concerns and personal health histories, as well as by the relative impact of heart disease, osteoporosis and cancer of the breast, colon and uterus on women's health in general. Doctors should advise women who choose not to take hormones of other ways to minimize heat disease and osteoporosis. Alternative approaches to protecting the heart include not smoking; following a regular exercise program; taking aspirin; and getting treatment for high blood pressure, high cholesterol and diabetes. Women can minimize bone loss through exercise, calcium intake and the judicious use of anti-osteoporotic medications. For many women, however, the potential benefits of hormone therapy on heart, bone, colon and quality of life will outweigh the risk of breast cancer. ❏

Questions

1. What is hormone replacement therapy?

2. What benefits are derived from hormone replacement therapy?

3. What cancers are being linked to the use of estrogen therapy?

Answers are at the back of the book.

48

With a growing shortage of human organs available for transplantation, the need to transplant animal organs and tissues into humans must now be considered. Foremost on everyone's mind is the risk to public health posed by xenotransplantation. In particular, what is the possibility of an emergence of "new" viruses and, with that, the risk of novel viral diseases that are not species-specific? These questions and many more are being addressed at the federal level by the CDC, NIH and FDA who are playing a lead role in developing guidelines and policies for xenotransplantation. Society must, along with each physician, scientist and patient, bear the responsibility for providing direction to this new field.

The Public Health Risk of Animal Organ and Tissue Transplantation into Humans

Frederick A. Murphy

Science, **August 9, 1996**

The many ethical, societal, and public policy issues surrounding animal organ and tissue transplantation into humans (xenotransplantation) have been expounded in a variety of settings over the past decade, but in the past year one particular issue has taken center stage—that of risk to the public health posed by novel viral diseases stemming from unique opportunities for species jumping. Underlying the issue is the shortage of human organs for transplantation and advances in immunological and surgical sciences that now promise the means to overcome cross-species rejection phenomena.[1] The argument over the level of societal risk presented by xenotransplantation has been intense and is far from being resolved. In hindsight the argument seems to have followed a roller-coaster course, alternating between a sense that the risk might be acceptable and a sense that it might not. For example, in 1994 the Food and Drug Administration (FDA) seemed headed toward approval of clinical trial protocols, but in 1995 such plans were suspended. In the past year two oversight groups, encouraged by FDA approval of a clinical trial involving the xenotransplantation of baboon bone marrow cells into a patient with acquired immunodeficiency syndrome (AIDS), have suggested that, with substantial safeguards in place, clinical trials should proceed.[2,3]

Focus is now turning to policy development to ensure that (i) the best scientific approaches are used to evaluate and quantify specific risks; (ii) comprehensive surveillance and virus screening, discovery, detection, and diagnostics systems are established; (iii) national clinical trial guidelines are made available to local institutional review boards; (iv) ample

Reprinted with permission from *Science,* Vol. 273, No. 5276, August 9, 1996, pp. 746-747.

communication takes place among involved professionals; (v) ethical concerns of patients and society are melded with scientific issues; and (vi) a permanent national oversight body is chartered. In this Policy Forum, factors pertaining to risk assessment are presented as bases for the kind of comprehensive policy development that must evolve over the next few years—hence this is a "work in progress," where new data will continually drive societal attitudes.

The viruses of xenograft donor species. There are about 4000 known virus species and about 30,000 strains and variants that infect humans, animals, plants, invertebrates, and microorganisms.[4] Although the risk posed by many viruses will require further evaluation, attention must be concentrated on viruses that are known to be pathogenic in donors or recipients and viruses with other suspected risk potential.

Known pathogenic viruses that might pose a risk in xenotransplantation include many adenoviruses, papovaviruses, papillomaviruses, parvoviruses, hepadnaviruses, morbilliviruses, filoviruses, hantaviruses, arenaviruses, arteriviruses, flaviviruses, and togaviruses. In evaluating the pathogenic potential of specific viruses, rather than whole categories such as the ones described, it will not be easy to determine which viruses represent a risk to the xenograft recipient alone, which represent a risk to society as a whole as a result of species jumping, and which may be dismissed as representing a minimal risk. An important aspect of policy development should be the construction of a list of the viruses of concern and an evaluation of the relative risk each poses in various xenotransplantation settings—this is a task that has not yet been done in a comprehensive way.

In particular, the risk associated with the presence in donor animals of certain retroviruses (including endogenous retroviruses, mammalian type C and D retroviruses, lentiviruses, and human T cell leukemia virus/bovine leukemia virus-like viruses) and certain animal herpesviruses (including herpes simplex-like viruses, Epstein-Barr-like viruses, cytomegaloviruses, and HHV6-, 7-, and 8-like viruses) must be considered further. Every potential donor species carries one or more herpesvirus, usually silently by a high proportion of individuals in the population and often capable of causing severe disease when infecting a heterologous species.

Sources of xenograft organs and tissues. Various sources have been used to obtain organs, tissues, and cells for xenotransplantation, including abattoirs, open colonies, closed colonies, specific pathogen-free (SPF) colonies, and gnotobiotic colonies. It is easy to say, "the 'cleaner' the donor animal the better," but there is need for data on the relative risk posed by various animal sources. Meanwhile, the commercial production of swine raised under SPF conditions and genetically engineered to lessen rejection when their organs are grafted into humans has been initiated by at least six biotechnology firms, and in the United Kingdom a national committee has been formed to draft a code of practice for using transgenically modified SPF swine.[5] Proposals for the development of SPF baboons are far less advanced, partly because of the time and cost involved in developing special rearing programs and facilities, and partly because of objections over the use of this species for this purpose.

What about the immunosuppression induced in xenograft recipients? Pathologically and pharmaceutically induced immunosuppression affects the escape from immune control of viruses already present in the body, such as cytomegaloviruses, papovaviruses, and papillomaviruses, and also favors persistent virus carriage and shedding. In such circumstances, mutations may continue to accumulate so that the virus population found late in the course of infection may be quite different from the original infecting virus. In most observations of this phenomenon, the virus shed late has been attenuated in pathogenic properties, but who is to say that this will always be the case? This is one of the most important issues facing policy developers and requires further research in animal models and in immunocompromised human patients.

What is the difference between a xenograft and an unsterilized biologic product derived from an animal and injected into a human? There are many examples of biologic materials derived from animals for use in humans; a number of these are subjected to viral-inactivation procedures, and many are regulated by the FDA, including porcine insulin (treated with HCl and ethyl alcohol), bovine thyroxin (treated with acid), porcine heart valves (treated with glut-araldehyde), bovine lung lipids (for treating hyaline membrane disease; treated with solvents), bovine adrenal cells (experimental, as a source of endorphins for long-term intractable pain relief; no treatment), and porcine skin (for burn repair; no treatment). In addition, fetal calf serum, calf serum, and horse serum are used in cell culture substrates for vaccine production and for many in vitro autologous cell manipulations. Such sera are heat-treated or γ-irradiated, but viruses occasionally survive such treatment. So, there may not be much difference between xenografts and unsterilized biologic products derived from animals. In fact, we have for many years been parenterally transferring some of the same kinds of viruses into humans that might now be considered a risk in the xenotransplantation setting. Such parenteral transfers, which to a large extent have proceeded without apparent harm, should nevertheless be reviewed in regard to lessons for policy development and should also be compared with experiences involving transfers of whole organs (allografts and xenografts, where cell-cell interfaces and microvasculature remain intact and eventually unite host and transplant).

Systems for viral discovery, detection, and diagnostics. One imperative of xenotransplantation policy development is the design of a national system for virus screening, discovery, detection, and diagnostics. This system could be applied to potential xenograft donor animals as well as to xenograft recipients and surgical staff. Despite our incredible power to diagnose viral diseases and detect viruses, it would be a mistake to think that methods in common use are all that sensitive and specific for the purposes at hand. This failing is mostly a result of the extreme compartmentalization of diagnostics technology and shortcomings in technology transfer and training. An infrastructural change, driven by national policy, is crucial to the development of the kind of labora-

tory resources that would meet public expectations.

In a national virology laboratory supporting leading clinical xenotransplantation centers, (i) the list of tests available for known viruses of concern would be comprehensive; (ii) the sensitivity of tests would be maximized; (iii) the specificity of tests would be adjusted to the purpose at hand (not too narrowly specific, such that variant viruses might be missed); (iv) the sampling of donor materials would be expansive and statistically sound; and (v) there would be a seamless cloth from standard methods through to avant-garde investigational methods. Of course, there are wide gaps between standard and investigational methods; the former are subject to quality control and are usually supported by reference laboratories and state and national reference centers, whereas the latter are research-driven and not subject to independent oversight. Tests for many viruses in xenograft donors would for some time have to be seen as investigational, in need of careful interpretation.[6]

There is a big difference between diagnostics and etiologic-agent searching. In the latter, many nonspecific approaches are used in complementary fashion (such as electron microscopy and evidence of viral growth in cell cultures inoculated with donor materials). Overall, despite the introduction of powerful molecular biologic methods (such as shotgun cloning, amplification by the polymerase chain reaction, representational difference analysis, and sequence-independent single-primer amplification), nonspecific methods are often poorly predictive of the presence of many kinds of viruses. Moreover, every application of such methods represents a major research project. The matter of how best to detect unknown viruses of potential concern in xenotransplantation must be regarded as a new field, open-ended and in need of greater research. This issue will be one of the linchpins of comprehensive policy development.

Understanding risk. One element that has helped bring about the national decision to allow further clinical trials has been an increasing public understanding of the concept of risk. In dealing with infectious diseases, the reality is that our best efforts may decrease but will never eliminate risk.[7] There is also a need to better understand the fundamental nature of the viral infection risks involved in xenotransplantation. Each virus of concern must be evaluated independently and quantitatively. Ultimately, risk may be revealed only through ongoing surveillance and clinical observation, but in complementary fashion, animal model studies may provide our best opportunity for understanding the mechanistic bases for species jumping. In such studies endogenous viral recombinants, complemented escape mutants, and other exotic, theoretical risks can be experimentally tested. Policy development must include a fundamental biomedical research base for clinical xenotransplantation sciences—a need that has not yet been comprehensively described.

The need for national leadership, coordination, and guidelines. At first, the question of risk associated with xenotransplantation focused on the individual recipient. This focus led to a consideration of many topics, including the nature of the virus, its pathogenesis, its pattern of transmission, and its stability. Prions—the agents of the spongiform encephalopathies—being so physically stable (for example, resistant to boiling, formaldehyde, and ultraviolet- and γ-irradiation), so insidious and persistent, and so difficult to detect in donors, seemed to represent the ultimate test of various risk management proposals and policy ideas.

In the past year, however, the question of risk has been expanded to cover the whole population that might come into contact with the xenograft recipient. The questions now asked include: What is the risk of novel viral diseases stemming from unique opportunities for species jumping? Will xeno-transplantation be the cause of the epidemic emergence of "new" viruses? Will the immunosuppression induced in xenograft patients amplify this risk? Here, another HIV-like virus/AIDS-like epidemic replaces prion diseases as the ultimate threat.

The answers to these population-based questions will come from many sources of expertise; however, to a greater extent than with individual health questions, the scientists at the national public health agencies—the Centers for Disease Control (CDC), FDA, and National Institutes of Health (NIH)—bear particular responsibility for providing direction. Their special expertise must be complemented with that from other areas, such as basic biomedicine, academic clinical medicine, veterinary medicine, laboratory animal medicine, and primatology.

At a recent workshop sponsored by the Institute of Medicine's Committee on Xenograft Transplantation, a strategy was developed to achieve national leadership, coordination, and guidance.[2] The strategy avoids regulation per se, calling instead for national guidelines to help local institutional review boards oversee clinical investigators. In keeping with a recommendation that the CDC, NIH, and FDA play the lead role in developing these guidelines, an interagency xenotransplantation working group has been formed, and draft guidelines are expected to be published in the Federal Register soon.[8] These guidelines will call for a national registry of patients and will describe in detail the kind of national surveillance and laboratory resources needed. The guidelines will reaffirm the principle that the issues at hand are societal in nature—they concern not just individual physicians, scientists, and patients—and will also reaffirm the need to continue to assess, manage, and communicate the risks involved.

References and Notes

1. F. Hoke, *Scientist* **10**, 11 (1995); R. E. Michler, *Emerging Infect. Dis.* **2**, 64 (1996).
2. CDC/FDA Xenotransplantation Working Group, *Guidelines for Xenotransplantation* (Federal Register, Washington, DC, in press); *Xenotransplantation: Science, Ethics and Public Policy* (Institute of Medicine, Committee on Xenograft Transplantation, Washington, DC, 1996).
3. L. K. Altman, New York Times, July 19, 1994, p. B6; December 15, 1994, pp. A1 and A16; December 16, 1994, p. A4; December 18, 1994, p. A5; December 19, 1995, p. B4; January 4, 1996, p. C19.
4. F. A. Murphy et al., *Virus Taxonomy: The Sixth Report of the International Committee on Taxonomy of Viruses* (Springer-Verlag, Vienna, 1995).
5. L. M. Fisher, *New York Times*, January 5,

1996, pp. Bus1 and Bus3; C. O'Brien, *Science* **271**, 1357 (1996) .

6. M. G. Michaels and R. L. Simmons, *Transplantation* **57**, 1 (1994) ; M. G. Michaels et al., ibid., p. 1462.

7. J. Lederberg, R. E. Shope, S. Oaks, *Emerging Microbial Threats* (National Academy of Sciences, Washington, DC, 1992).

8. L. E. Chapman et al., *N. Engl. J. Med.* **333**, 1498 (1995).

9. I thank L. E. Chapman and T. M. Folks (Retrovirus Diseases Branch, National Center for Infectious Diseases, Centers for Disease Control and Prevention, Atlanta) for their generous advice. ❑

Questions

1. Why must animals raised under specific pathogen-free conditions be genetically engineered before they are used as xenograft donors?

2. Why must better virus screening methods be designed?

3. What biological products derived from animals have been designed for use in humans?

Answers are at the back of the book.

49

Irving Weissman's discovery has hit the scientific community like a hurricane leaving behind in its path a whirl of excitement and controversy. It all began when he staked his claim in the form of a patent on finding a human hematopoietic stem cell, a blood-forming cell capable of generating an endless supply of red cells, white cells, and platelets. Imagine the potential of such a monumental achievement as replacing a patient's entire blood supply in the treatment of cancer, AIDS, and inherited diseases. The question is should one man and his colleagues be the proud owners of a living cell? Some of the scientific community believes otherwise since much of the work had already been done on other laboratories. Whatever the case, Weissman's patent has cast much interest and money into the field of stem cells and it stands to make big money for the winners.

The Mother of All Blood Cells

Peter Radetsky

Discover, March 1995

Stem cells, capable of generating an endless supply of red cells, white cells, and platelets, have also generated a heated scientific controversy—and millions of dollars for the man who claims to have found them.

Deep in the very marrow of our bones reside the living forebears of our blood, the hematopoietic ("blood forming") stem cells. From these rare and elusive pluripotent cells arise our oxygen-carrying red blood cells, the tiny platelets that facilitate coagulation, and the disease-fighting white cells of our immune system. The hematopoietic stem cells are nothing less than the springs that feed the river of life that flows through our veins. And Irving Weissman has found them.

To be precise, Weissman, a Stanford immunologist, and his collaborators at SyStemix, the Palo Alto, California, biotech company he cofounded in 1988, claim to have found a strong "candidate" for

these remarkable cells. But they're not fooling anyone. So confident are they, that SyStemix has patented not only the process used to find the cells but the cells themselves, in effect claiming ownership of these biological entities. So confident is the giant Swiss drug-and-chemical company Sandoz Ltd., that it has bought 60 percent of the SyStemix stock for a reported $392 million, making the 55-year-old Weissman and his stockholders instant millionaires.

The stakes are indeed high. In addition to stem cells' importance in basic research, they may make possible a host of breakthrough medical advances. By giving patients the ability to make an entirely new blood supply essentially from scratch—and thus the ability to regenerate key components of the immune system—stem cell therapies could result in new treatments for various cancers, allow for bone marrow transplants without the need for rejection-fighting drugs, make possible powerful strategies against AIDS and other blood infections, and provide genetically engineered antidotes to a wide range of inherited diseases. No wonder Weissman has caused such a stir.

Yet none of the pioneering researchers in these fields currently employ the Weissman recipe for isolating stem cells—except, of course, those at SyStemix. Inarguably, the hunt for the stem cell preceded Weissman, and some say it would be proceeding just fine without him. While researchers agree that Weissman has brought recognition to a previously little-known field, they disagree as to the value of his scientific contribution. Some consider his work pivotal, others merely useful, others virtually irrelevant—regarding him, in Weissman's own words, as "a snake oil salesman." Despite Weissman's controversial patent, there is continuing debate as to exactly what stem cells really are and how they might best be tracked down. Says stem cell pioneer and Weissman competitor Malcolm Moore of the Memorial Sloan-Kettering Cancer Center in New York City, "No one has yet definitively isolated a stem cell. There's been a lot of talk, but it hasn't yet been pinned down so we can say, '*This* is the stem cell.' "

• • •

The modern search for the hematopoietic stem cell began with the detonation of the atomic bombs over Japan in 1945. Researchers could easily see that the intense blasts of radiation destroyed blood cells and that people often died within weeks of exposure. But scientists mimicking that exposure in mice soon realized that these deadly effects could be prevented by transplanting bone marrow from genetically identical donors into the irradiated mice. The injected marrow revived the irradiated blood; thus, the researchers reasoned, the marrow must contain cells capable of regenerating other blood cells, something that mature blood cells are not able to do.

In the 1960s those speculations became fact. James Till and Ernest McCulloch of the Ontario Cancer Institute in Toronto found that after bone marrow cells were injected into irradiated mice, the animals developed nodules on the spleen. Each nodule was chock-full of white and red blood cells. By tracking genetic markers in the cells' chromosomes,

Till and McCulloch saw that the cells within each nodule had all derived from a single progenitor: one per nodule. Then, by simply counting nodules, they were able to estimate the number of progenitor cells in each batch of transplanted marrow. The cells turned out to be rare, about 1 in 1,000. Furthermore, Till and McCulloch found that in addition to generating a wide range of new blood cells, these progenitor cells were also able to reproduce themselves.

Based on this evidence, Till and McCulloch came up with a scenario that has been considered gospel ever since. All blood cells, they said, arise from a few hematopoietic stem cells, which are hidden away in the bone marrow. (Blood-forming stem cells are not the only ones we harbor: there are supposedly stem cells for skin, liver, and the intestines, as well as stem cells behind the generation of eggs and sperm.) These cells, as remarkable as they are rare, can both renew themselves and produce trillions upon trillions of blood cells, an inexhaustible supply for the life of their host body. When these new cells mature and die off—human red blood cells, for example, last only 120 days—the stem cells produce more to take their place. On average these cells produce an ounce of new blood—some 260 billion new cells—each and every day.

Irv Weissman was a Stanford medical student and research associate when Till and McCulloch did their groundbreaking work. "I knew those experiments," he says. "I knew them cold. They were thrilling." The experiments gave him some insight into the problems of organ transplantation, something he'd been interested in since his high school years in Great Falls, Montana. "I thought that the most important thing would be to understand the development of the immune system," he says. "If you understood that, then when you did transplants you'd know what was going on." Till and McCulloch's stem cell revelations gave him a path to follow. "I began getting further and further into understanding white cell development, moving backward from mature cells to earlier and earlier cells."

Of course, Weissman wasn't the only one moving backward. Researchers all over the world were beginning to look at the early, immature blood cells. Among them were the hematologists and biologists who made up the "Dutch Mafia," as biophysicist Jan Visser laughingly describes himself and his compatriots. And it was the Dutch Mafia that first hit stem cell pay dirt. In 1984 Visser announced that he and his colleagues in the Netherlands had isolated stem cells in mice.

It was a startling achievement, one that had eluded even Till and McCulloch. The Ontario researchers had only been able to document the cells' existence, not to pin them down. But in the intervening years molecular techniques had grown more sophisticated. Visser had at his disposal molecular probes designed to find their prey by homing in on any number of unique characteristics. The task of finding the theorized stem cells among all the varied cells in a sample of marrow was therefore akin to trying to pick someone out of a crowd by looking for a particular combination of hair color, weight, and nose shape. This didn't mean the job was an easy one, however. Stem cells are comparatively primitive cells that lack the diversity of features associated with their mature progeny; they seem to be distinguished primarily by their dearth of unique characteristics.

To tease out the stem cells, then, Visser employed a three-pronged strategy. First he separated the cells by density. It had been discovered some years earlier that cells with stem cell activity—that is, cells that gave every indication of indeed being the long-sought hematopoietic cells—tend to be lower in density than other bone marrow cells, so Visser placed a batch of bone marrow cells (between 60 million and 100 million of them) in a centrifuge and culled only the ones that rose to the surface. That one step got rid of some 90 percent of all the cells.

Next he turned to a substance commonly used in laboratories to purify proteins: wheat germ agglutinin. Agglutinin fuses with certain sugars associated with proteins, and Visser had found that it also sticks to the sugars in the membranes of cells with stem cell activity. So he took the remaining 10 percent of the cells and mixed them with wheat germ agglutinin

tagged with a fluorescent dye. With the help of a fluorescence-activated cell sorter, he was able to separate out only those cells that emitted a fluorescent glow, a sign that the tagged agglutinin was holding tight. In this way Visser further reduced his sample by 90 percent. He was now left with just 1 percent of his original bone marrow mix.

Finally he employed monoclonal antibodies. Antibodies are large Y-shaped molecules that are among the immune system's prime infection fighters. They make a beeline for foreign proteins, grab them, and mark them for destruction by other immune forces. By the early 1980s these tiny guided missiles were among the favorite tools of molecular biologists, since they could be engineered to go after almost any target a scientist might choose. Visser sent them after one of the few stem cell characteristics then known: a protein called H-2K, which he had discovered in greater numbers on the surfaces of stem cells than on any other cell. Cells the antibodies ignored couldn't have the protein and thus couldn't be stem cells. These he cast aside.

The result was a further narrowing of the search. The antibodies reduced the number of cells by two-thirds: what remained was just three-thousandths of the original blend, only about 200,000 cells. When Visser injected these relatively few cells into irradiated mice, he found that it took no more than 200 of them to regenerate each animal's entire blood system. In other words, there was at least 1 stem cell in every 1,000 bone marrow cells, the very proportion Till and McCulloch had come up with. Of course, Visser's sample wasn't pure—there were other cells in the mix—but out of an initial crowd of tens of millions, he was pretty close. Three years later, using newer sorting techniques, he whittled it down even further, using only 30 cells to save an irradiated mouse. He now estimates that stem cells represent 1 in 10,000 marrow cells.

Visser published his original results in the *Journal of Experimental Medicine* in 1984. Four years later Weissman announced in the journal Science that he and his colleagues had found the mouse stem cell. Whereas Visser's work had elicited polite praise, Weissman's made headlines. "*The Journal of Experi-*

mental Medicine is considered scientifically one of the best journals," Visser notes with some irony. "*Science* is more popular."

Weissman and his team had taken a much more narrow approach to ferreting out their prey: they relied solely on a variety of monoclonal antibodies, each designed to pick out a different stem cell surface protein—or proteins on other cells that they had found to be absent on stem cells. For instance, one group of monoclonal antibodies targeted proteins found only on the surface of mature bone marrow cells, thus allowing the researchers to get rid of almost everything that was not a stem cell. Weissman dubbed the remaining cells Lin- ("lineage minus") because the antibodies had subtracted all other cell lineages. "We used those antibodies to get rid of 90 percent of the bone marrow," he says. "The 10 percent that was left had stem cell activity."

To pinpoint the source of that activity more precisely, they used two other monoclonal antibodies targeted to two surface proteins they knew appeared on stem cells. One—Thy1—is found in low concentrations (designated "lo"); the other—Sca1—is far more common (designated "+"). By throwing out the cells that did not display Thy1 or Sca1, Weissman brought the number of remaining cells down to just .05 percent of the whole. He dubbed the cells left behind Thy1^{lo}Lin$^-$Sca1$^+$; like Visser, he found that only 30 of them were needed to reconstitute the full range of blood cells in an irradiated mouse. What he had was a "virtually pure" batch of stem cells.

• • •

Reaction was strong and swift. *Science* accompanied Weissman's paper with a news story entitled "Blood-Forming Stem Cells Purified," in which Weissman, without so much as a word acknowledging earlier work toward the same goal, was quoted as saying, "This is the end of the particular road that was the search for the stem cell." He contended that his methodology—based on identifying the actual look of the cells—was so efficient that it might be used to go after human stem cells. Neither Visser—who had focused more on the cells' density and their propensity to bind to agglutinin—nor anyone else had made a claim like that.

SCIENTISTS CLOSE IN ON A VITAL BLOOD CELL, proclaimed the *Wall Street Journal.* THRILLED TO THE MARROW: BIOLOGISTS FINALLY CORNER THE RARE FOREBEAR OF ALL BLOOD CELLS , announced *Scientific American.* Weissman spread the word over television and radio, and the news was covered worldwide by the Voice of America. But despite the public applause, not all the response from the scientific community was complimentary. An editorial in *Immunology Today,* for example, while acknowledging that the researchers had indeed isolated a cell population with stem cell activity, asked, "But does this represent any advance on previously published data?" The editorial pointed out that Visser's work had "generated populations with similar characteristics and only moderately less purity," and concluded, guardedly, "We have not yet reached the end of the road; perhaps just one of the side streets."

The suspicion spread that Weissman had done little that was new but had nevertheless claimed credit for a monumental discovery. In the process he had slighted the achievements of earlier pioneers, Visser's in particular. Whether Weissman's greater recognition was the result of his being published in the right place at the right time, or his being American rather than Dutch, or his simply knowing his way around a press conference better than Visser, many scientists felt that Weissman was receiving attention out of proportion to his accomplishment.

"Did the Weissman group make a real contribution in the mouse? No, not at all," declares Malcolm Moore. "All of the work had already been done elsewhere. Visser had done it a long time before. So there were some people very, very angry that he had taken credit for discovering the stem cell when he hadn't done any of the primary work."

"It's unfortunate that it worked out that way," says stem cell researcher Ihor Lemischka of Princeton. "I think Weissman would be the first to admit that his getting all the credit was unfair. It caused a lot of bad feeling. Now there are two camps in the field. There's the Dutch axis and the Weissman crowd, and everybody else is in between."

Weissman has since explicitly acknowledged Visser's prior contribution. He didn't do so at the time, he says, because he simply wasn't aware of

Visser's work. "I didn't even know about it until we finished our work," Weissman claims. "Early in my career a prominent scientist told me that when he started something in earnest, he quit reading the literature. He didn't want to be distracted by what others were doing, or spoil the fun of discovery. I use that as an excuse. It's a bad thing to admit this, I know."

Whatever its cause, Weissman's omission had unfortunate consequences. Says Moore, "Weissman was an immunologist who suddenly hit the interface with experimental hematology without appreciating that people had been working for many, many years and knew about all these things. But somehow he obtained the credit for discovering stem cells. He trod on many toes."

Ten of those toes belonged to Visser, today the head of the recently formed stem cell laboratory at the New York Blood Center in Manhattan. Yet Visser is far less strident in his assessment of Weissman's tactics than are some of his colleagues. In fact, he seems almost to admire Weissman's flair. "Did I feel that Weissman stole my thunder?" he muses. "At first I did. I felt sort of sorry about it. But that soon went away. He raised attention to this field that I could never have raised. In his jet stream, in all the noise he makes, I was drawn with him. We both traveled to meetings around the world to explain our differences. No, after all it was no problem. It was good for me. It gave me a lot of attention."

Whether he was the first to find stem cells, it is undeniable that Weissman devised a very useful approach. It was precisely because he was an outsider that he was able to pick and choose among existing techniques and combine them in a fresh way. $Thy1^{lo}Lin^-Sca1^+$ was an original recipe for a stem cell.

"The real value of the Weissman purification is that it describes cell-surface differences," says Lemischka. "Thy1, Sca1, and lineage markers—these are tangible molecules. So in terms of saying, 'A stem cell looks like this, it has on its surface this and that, but not these,' Weissman defined the stem cell for the first time. Weissman's accomplishment is more than just a purification like Visser's; it is really informative."

McCulloch is even stronger in his praise. "Weissman's remains the most extensive purification that's been achieved," he says. "I consider his work to be seminal. The application to man was just a step beyond what he did in the mouse."

• • •

All the same, it was quite a momentous step. By the winter of 1991 Weissman and his colleagues at SyStemix had adapted the mouse stem cell recipe and announced that they had found the human stem cell. As with the mice, they selected against the lineage markers, and for the Thy1 marker. Sca1 wasn't a practical marker to use in human cells—in humans the gene that encodes the protein has many similar-looking relatives—so the researchers substituted a protein called CD34, which had been discovered some years earlier. They dubbed the human stem cell $Thy1^+Lin^-CD34^+$. Once again Weissman's individual ingredients weren't new, but his combination was. Yet once again the scientific community wasn't overawed—other researchers, using many of the same markers, were already finding similar cells.

On the other hand, none of those other researchers announced their achievement the way Weissman did—with a patent. In a dramatic departure from normal scientific procedure, Weissman and his colleagues patented their method for finding their purported human stem cell *and* the stem cell itself some five months before their work appeared in the April 1992 issue of the *Proceedings of the National Academy of Sciences*. Since a patent confers exclusive rights to the thing patented, this meant that Weissman and his colleagues were the proud owners of a living human cell.

It was an audacious stroke, if not an entirely unprecedented one. "You know, people patent body components all the time," says Weissman. He's right. Genes, growth factors, blood proteins—all have received patents. Even CD34 is patented—SyStemix had to pay to use the protein. But an entire living cell?

"Next thing you know, someone will patent a zygote, and you won't be able to have a baby without a license," quipped Sloan-Kettering hematologist David Golde at the time of the announcement.

"If somebody wants to commercialize having babies, who knows?" replies Weissman. As with any patent, SyStemix's patent is specifically designed to forestall the commercial exploitation of stem cells by anyone other than itself. And the company intends to guard its privilege aggressively. "It doesn't matter what process people use, they will always be infringing," declared SyStemix's then president, Linda Sonntag.

Such talk infuriates Moore and others in the field. "It seems to me an absurd patent," Moore says. "I predict it won't withstand any challenge. If they think that anybody who tries to separate and grow stem cells is going to have to pay a licensing fee to SyStemix, they've got another think coming. Because people can separate stem cells using different criteria from the ones outlined by the SyStemix patent." For example, in January, Harvard Medical School researchers announced that they had developed an entirely novel and relatively simple method of isolating stem cells based not on antibodies but on the principle that stem cells, which spend most of their time in a quiescent, nondividing "deep sleep," are relatively unresponsive to proteins known as growth factors. The stem cells' progeny, on the other hand, do respond to growth factors. So the Harvard team, led by hematologist David Scadden, turned on the stem cells' offspring by applying those factors and then promptly killed them by smothering the same cells with an anticancer drug called 5-FU, which attacks metabolically active cells. Left behind were stem cells.

Jan Visser is much more philosophical than Moore. Visser sees the move as a practical necessity. "Weissman needed the patent for his company."

It's an analysis with which Weissman agrees. "Money people won't invest in something that's not patented," he says. "And if you don't get money, you can't do big-time research."

Weissman can do that research. Almost immediately after the patent announcement, Sandoz bought controlling interest in SyStemix, ballooning Weissman's net worth, at the time, to an estimated $24 million. "It hasn't changed my life," he says, then chuckles. "Well, I drink better wines than I did."

"I would have liked to be as rich as he is," says Visser wistfully. "It's my own mistake. He did the right thing—from a family point of view." He laughs. "My children are complaining."

• • •

In the clinical realm, however, the SyStemix patent has yet to become a real factor. In its own labs, SyStemix has shown that its human stem cells can indeed regenerate blood in mice that have been bred with a human immune system—the so-called SCID-hu mice. And the company has received permission to begin testing its ability to do so in humans by injecting the cells into 20 multiple myeloma patients whose immune systems have been ravaged by chemotherapy. These tests should be starting right around now; it's taken this long to develop a technology that can use the $Thy1^+Lin^-CD34^+$ recipe to isolate stem cells in large enough quantities to use in humans.

But none of the other clinical trials going on right now use the Weissman recipe. At the New York Blood Center, immunogeneticist Pablo Rubinstein is spearheading a worldwide effort to transplant stem cells by utilizing human neonatal cord blood. His approach is simplicity itself. In the fetus, stem cells continue flowing through the blood before settling into the bone marrow a few days after birth. Cord blood is therefore relatively rich in stem cells; Rubinstein and his colleagues can simply collect and transplant that whole blood and know that they are transplanting stem cells. Since the fall of 1993 the team has helped perform transplants on 13 patients—all but one of them children—suffering advanced stages of leukemia or inherited disease. The idea is to provide these people with a new source of blood to replace their own diseased cells. The results have given researchers grounds for hope. Although 3 of the patients (including the adult) died from unrelated complications, the other 10 are doing well. Their blood has been fully regenerated, their disease put into remission.

Moreover, none of these patients experienced the bane of transplantation efforts, graft-versus-host disease, in which the introduced cells recognize the host's body as foreign and attack it. "With pure stem cells, there should be no graft-versus-host disease," says Rubinstein. "The immune system generated from such a cell would mature in the recipient and thus would be trained to recognize the recipient as self."

Another trial using cord blood began at Children's Hospital in Los Angeles in the spring of 1993, when pediatric gene therapist Donald Kohn performed stem cell gene therapy on three newborns afflicted with an inherited immune system disorder called ADA deficiency, known more popularly as the bubble boy disease. Children with the disease lack the gene to produce the enzyme adenosine deaminase, or ADA. Without ADA the immune system cannot function; such children face certain early death.

ADA deficiency was the target of the very first gene therapy trials, begun in 1990 at the National Institutes of Health. Those trials involved putting corrected genes into mature white cells rather than stem cells. The corrected cells thus eventually die, which means that the therapy—while revolutionary and successful—will be able only to curtail, not cure, the disease. Stem cells, on the other hand, might provide a true cure. In his attempt to provide that cure, Kohn actually needed to isolate the stem cells in the babies' cord blood: he did so by targeting the CD34 protein, inserting the ADA gene in the cells, and returning them to the infants. He's hopeful that the genes have taken up residence in the stem cells and are being packaged into their progeny. "For the first few months we didn't see any cells with the gene," he says. "Then we started seeing a few, maybe 1 in 10,000. Now we're up to about 1 in 1,000." Although in the meantime the infants are receiving injected doses of ADA enzyme, Kohn's hope is that stem-cell-generated cells containing the inserted gene will eventually produce enough enzyme to obviate the need for supplemental therapy and effectively cure the children.

These trailblazing efforts may presage a spectacular future for stem cell therapies. Malcolm Moore and his team are inserting drug-resistant genes into stem cells identified by CD34 and other markers so that they can give the cells to leukemia patients or cancer patients undergoing chemotherapy, which tends to destroy healthy white cells along with the diseased cells. A number of groups—including one at SyStemix—are looking at inserting anti-HIV genes into stem cells as a possible treatment for AIDS. Finally, though it's a long shot, the ability to isolate stem cells may even lessen the escalating need for blood transfusions.

In the end, controversy and hard feelings aside, it seems that Weissman has sparked a revolution. Today there is interest in stem cells, and money available to fuel that interest, as never before. "Because of the public relations work that Weissman did, it became easier to get grants," says Visser. "Because Sandoz bought SyStemix, all the big companies started to look at stem cells."

"What did Irv Weissman bring?" asks stem cell researcher Norman Iscove of the Ontario Cancer Institute. "He made the field live. He brought it into headlines, into newscasts."

No one appreciates the field more than Weissman. "I think you'll see the first practical large-scale use of stem cells in three years for cancers, five to seven years for other conditions," he says. "The great thing is that we have developed this population of cells that are just the right thing. They're not like drugs, which have side effects. You know that what you're providing is the right thing because stem cells are the product of over a billion years of evolution."

And Irv Weissman owns them. ❏

Questions

1. What are stem cells?

2. How did Irving Weissman discover stem cells?

3. How can stem cells be used for therapy?

Answers are at the back of the book.

Answers

SECTION ONE: LIFE IN THE CELL

1. Conversations in a Cell

1. In signal transduction, protein molecules combine on the cell surface to elicit a response that ends in the nucleus activating or deactivating genetic mechanisms.
2. Proteins are long strings of amino acids with side chains that can rotate and adopt different structures.
3. Specific genes could be activated when they are needed and deactivated when harmful.

2. Epithelial Sodium Channels: Their Role in Disease

1. It is a channel responsible for the electrogenic absorption of sodium.
2. These channels are found in the cortical collecting duct of the kidney, the distal colon, and the lungs.
3. Liddle's syndrome is a monogenic, salt-sensitive form of hypertension.

3. Flexing Muscle with Just One Amino Acid

1. Troponin-C is a protein that changes shape in response to calcium, allowing other proteins to interact to induce muscle contraction.
2. Glutamate, a negatively charged amino acid, is attracted to the positively charged calcium causing them to move closer in the

protein.
3. NMR determines the position of atoms within a protein from the way they resonate in a magnetic field.

4. Hemochromatosis: The Genetic Disorder of the Twenty-first Century

1. Hemochromatosis is a condition of abnormal iron absorption across the gut epithelium.
2. In the average diet, 15-20 mg of iron is available, however, only 1-2 mg is absorbed.
3. The primary treatment is phlebotomy, a removal of the blood, on a weekly basis until iron stores are exhausted.

5. New Hope for People with Sickle Cell Anemia

1. Sickle cell anemia is characterized by an abnormal type of hemoglobin which causes red blood cells to become hard and sickle shaped.
2. In sickle cell anemia, the blood flow can be interrupted to any of the major organs, causing severe pain and organ damage.
3. The body destroys the abnormal red blood cells causing a type of anemia.

6. Ringing Necks with Dynamin

1. Synaptic vesicles, found in the synaptic knob of neurons, are pouches that store

specific neurotransmitters for release each time a neuron fires.

2. Endocytosis is the process of internalization of extracellular material within a cell.

3. Dynamin is a large protein whose helical structure wraps around the neck of membrane vesicle and helps to pinch them off from the plasma membrane.

7. Revisiting the Fluid Mosaic Model of Membranes

1. The fluid mosaic model describes the plasma membrane as a phospholipid bilayer with proteins associated with either the inside or outside spanning the membrane completely in an unrestricted manner.

2. Proteins confined transiently within the membrane are the cell adhesion molecules, neural adhesion molecules and epidermal growth factor receptors.

3. The membrane-spanning proteins are sterically confined to the cytoplasmic face of the membrane face based on their cytoskeletal mesh size.

8. Biocarbonate Briefly CO_2-Free

1. Carbon dioxide and water form carbonic acid, which dissociates into H^+ and bicarbonate.

2. Carbonic anhydrase catalyzes the reaction of CO_2 with water.

3. Potassium creates a gradient necessary for the cotransport of bicarbonate in this new system.

SECTION TWO: WHAT'S NEW IN TISSUE ENGINEERING

9. Paralysis Lost

1. Nerve axons have the ability to grow from the spinal cord to the muscle.

2. Axons grow about 1 millimeter per day.

3. Neurotrophic factors help stimulate axon regrowth.

10. Tissue Sealants: Current Status, Future Potential

1. Fibrin is made of fibrinogen and thrombin which promote clot formation in blood coagulation.

2. Careful donor selection and cryoprecipitation of blood plasma are methods used to ensure viral safety.

3. Fibrin sealants could be used as a delivery system for drugs and tissue growth factors.

11. Making the Connections in Nerve Regeneration

1. Injuries to the brain or spinal cord could benefit from central nerve regeneration.

2. To promote survival and regeneration of peripheral neurons, they need a full complement of basal lamina of Schwann cell, extracellular matrix and growth factors.

3. Some factors which may be inhibitory to central neurons are myelin-associated glycoproteins, the presence of peripheral nerve myelin and gray matter from the brain.

12. Designer Tissues Take Hold

1. Scientists synthesize a scaffold with biocompatible materials to attract a particular cell type.

2. For a synthetic material to be safe, it must not conflict with the body's immune system.

3. Hip joints and dental implants are two applications of hybrid technology.

13. Researchers Broaden the Attack on Parkinson's Disease

1. Low survival of the transplanted tissue accounts for some of the variability.

2. It would produce almost pure dopaminergic neuroblasts and reduce or eliminate the dependence on fetuses.

3. It is a protein that does not cross the blood-brain barrier.

14. Fetal Attraction

1. Substantia nigra, the region that supplies the brain with the neurotransmitter dopamine, is destroyed.

2. L-dopa, or levodopa, is a drug that can pass through the blood-brain barrier to be converted to dopamine.

3. Transplants from the patient's own tissue.

SECTION THREE: THE BRAIN, THE NERVOUS SYSTEM, AND THE PROPERTIES OF NEURONS THAT SERVE THEM

15. Racked with Pain
1. The three types of pain are protective pain, reparative pain and enigmatic pain.
2. It is difficult to diagnose and treat because it has no obvious cause and cannot be alleviated with drug therapy.
3. Damage to sensory nerves coming from the body surface diverts touch and pressure messages into pain pathways in the spinal chord.

16. Revealing the Brain's Secrets
1. The gene codes for a protein that appears to contribute to the premature death of certain neurons in the brain.
2. Neuroprotective drugs are being developed by scientists to guard brain cells against damage and death and even help them to regenerate.
3. Neurotrophic growth factors are small proteins produced by the brain which may help keep them alive during conditions of disease and injury.

17. Action Potentials in Dendrites: Do They Convey a Message?
1. These are action potentials that propagate into the dendritic tree away from the cell body.
2. The dendritic action potentials are broader than somatic action potentials and their amplitudes decrease with increasing distance from the cell body.
3. The neuron has the capability to convey information about its level of activity to pre- and postsynaptic sites.

18. Brain, Heal Thyself
1. Stem cells replace injured and dead cells.
2. No one has conclusively isolated stem cells in the adult mammal brain, but scientists have induced mouse brain cells to act like stem cells in the lab.
3. To repair the injured brain, genetically engineered stem cells can be injected into the area to deliver and spread a missing enzyme.

19. What Is Leptin for, and Does It Act on the Brain?
1. Leptin is a hormone secreted by adipose cells.
2. Leptin causes weight loss in the body.
3. Scientists used a leptin-deficiency model of human obesity.

20. Dendrites Shed Their Dull Image
1. Each synaptic potential adds up with all the others moving through the dendrites, which if great enough, will result in an action potential.
2. In the patch-clamp technique, electrodes are pressed against the neuron's outer membrane to form a tight electrical seal so that action potentials can be measured at the dendrite.
3. Backward-traveling of signals can strengthen the synapse (plasticity), which will allow it to produce a better response to future signals.

21. New Clues to Brain Dopamine Control, Cocaine Addiction
1. Neurotransmitters relay signals between nerve cells in the brain.
2. Dopamine is essential in controlling movement, cognition and emotion.
3. A transporter is responsible for removing the excess dopamine from the synapse.

SECTION FOUR: THE MECHANICAL FORCES AND CHEMISTRY OF THE SENSES

22. NO in the Nose
1. Nitric oxide binds to the enzymes of bacteria stopping their growth and reproduction.
2. Nitric oxide is a potent vasodilator of blood vessels.

3. Nitric oxide from nasal air is high enough in concentration that it can be pumped to the lung through a ventilator to help improve blood pressure.

23. Lasers Win Out in Glaucoma Trial
1. Glaucoma is a disease which is caused by intraocular fluid build-up behind the iris.
2. The normal pressure in the eye is around 2000 pascals.
3. Laser treatment of the trabecular meshwork changes the matrix of the cells allowing them to become more permeable to the flow of intraocular fluid.

24. Stimulating Hair Cell Regeneration: On a Wing and a Prayer
1. Hearing loss is due to the death of hair cells, the receptor cells in the cochlea of the inner ear.
2. Scientists are studying birds to help answer questions about mammalian hair cell regeneration.
3. In mature mammals, no ongoing cell division in the auditory orvestibular sensory epithelia occurs, however, some low level of mitotic activity in the vestibular epithelia is present suggesting that production of new cells is possible.

25. I Spy with My Moulded Eye
1. Using corneal topographic analyzers, the contours of the cornea can be mapped and then redesigned to correct any defects.
2. Short-sighted people would wear molds that would flatted their corneas to help focus the light on the retina instead of in front of it.
3. Long-sighted people would wear molds that help round the cornea which brings the focus onto the retina from behind the eye.

26. Mechanoreceptive Membrane Channels
1. Phasic and tonic receptors
2. Mechanical stimulation leads to a transient increase in a receptor membrane's conductance to the ions sodium, potassium, and sometimes calcium.
3. Certain drugs can block the channels for study in various types of cells.

27. The Smell Files
1. Within the olfactory bulbs are tiny glomeruli that act as neural junctions to sort the signals coming in from the receptors about odor components.
2. Signals don't get mixed up because only neurons bearing one kind of receptor converge on a single glomerulus.
3. An odor is distinguished by the pattern of glomeruli it activates.

28. The Sense of Taste
1. Salt, sour, sweet, and bitter
2. Taste buds
3. G-proteins

29. A Taste of Things to Come
1. Salty, sour, sweet, and bitter
2. Each taste is mediated in some way by signaling pathways that result in receptor cell depolarization, release of neurotransmitter, and synaptic activity.
3. A Ga subunit that is highly specific to taste neurons and used in the bitter signaling cascade.

SECTION FIVE: IMPORTANT MESSENGERS OF THE RENAL, ENDOCRINE, RESPIRATORY, CARDIOVASCULAR, AND REPRODUCTIVE SYSTEMS

30. Hemoglobin Reveals New Role as Blood Pressure Regulator
1. Hemoglobin, found in red blood cells, gives up oxygen and picks up carbon dioxide at oxygen-poor tissues.
2. Nitric oxide is produced in the endothelial cells that line the blood vessels.
3. Hemoglobin scavenges for nitric oxide and maintains a balance to ensure that blood vessels are kept open.

31. Cardiac Performance: Growth Hormone Enters the Race

1. Growth hormone's primary role is in cardiac development.
2. Growth hormone can increase myocardial mass and enhance contractility with insulin-like growth factor-1 acting as its local effector.
3. They could be used under certain conditions to prevent a decline in skeletal muscle and cardiac muscle performance that comes with advancing age, catabolic disease and chronic illnesses.

32. High Blood Pressure

1. Blood pressure is the force exerted against the walls of the arteries and arterioles which carry blood from the heart.
2. The risk of hypertension increases with age as arteries lose elasticity and become less able to relax.
3. Diuretics lower pressure by eliminating sodium and water, thereby reducing the volume of blood in circulation.

33. Thyroid Diseases

1. The thyroid regulates the body's basal metabolism.
2. Thyroid-stimulating hormone, released from the pituitary, signals the thyroid to produce more T3 and T4 when needed.
3. Hypothyroidism

34. Assisted Reproduction

1. FSH, follicle-stimulating hormone, is produced by the pituitary gland in an effort to get the ovary to release an egg.
2. 10-15%
3. All have similar pregnancy rates, averaging 18-19%, but vary between clinics.

35. Not by Testosterone Alone

1. SRY gene
2. MIS actively suppresses the development of the female reproductive tract.
3. The MIS gene is found on the X chromosome so both males and females have it.

36. Exercise-Induced Renal and Electrolyte Changes: Minimizing the Risks

1. Plasma volume and electrolyte stores are depleted due to the loss of water and sodium in sweat.
2. Sweat loss can exceed 2 L./hr.
3. The primary ingredients are water and sugar in the form of corn syrup. Some sodium and other electrolytes can be found, as well as water-soluble vitamins.

SECTION SIX: HEALTH AND THE ENVIRONMENT

37. Can You Catch a Heart Attack?

1. Antibodies to *C.pneumoniae* were found in the blood of patients with coronary heart disease and its protein and DNA in the atheroma of diseased tissue that blocks the coronary arteries.
2. *C. pneumoniae* travels with macrophages through the lung's alveoli, enters the bloodstream, and then eventually rests on a damaged blood vessel wall to cause an infection.
3. Improvements in lifestyle are a major reason for the decline in death rates from heart disease, as is the use of broad spectrum antibiotics such as tetracycline which kill *C. pneumoniae.*

38. Skin-Deep Stress

1. Some stress may help to redistribute white blood cells in the body and not to destroy them as once thought.
2. They concentrate in the skin and other locations of the immune system such as the lymph nodes and bone marrow where they can be ready for an immune challenge.
3. Corticosterone is released from the adrenal glands during stress.

39. Hormone Hell

1. These chemicals are believed to disrupt animal hormonal systems causing an array of unexpected effects by imitating natural

hormones.

2. Developing fetuses—of any species—are sensitive to these chemicals because their exposure is many times greater than the daily adult exposure.

3. Synthetic chemicals tend to accumulate in the human body while natural chemicals are quickly metabolized and excreted.

40. Science Tracks Down the Training Dangers

1. Athletes have made considerable improvements in sporting achievements over recent decades due to more demanding training regimes.

2. Young female gymnasts, whose bones are still growing, display four times the occurrence of fractures in the vertebrae of the lower back than the normal population.

3. During exercise, muscle tissue uses glutamate which is essential for the metabolism of lymphocytes.

41. Could Women Take a Lead Over Men in the Long Run?

1. The male hormone testosterone increases muscle mass while the female hormone estrogen increases body fat.

2. Women take longer to deplete their glycogen stores.

3. Endurance training raises the consumption of free fatty acids in the body which can be used instead of glycogen.

42. Beyond Steroids

1. They are using the drug as a potent anabolic agent to increase muscle mass and strength.

2. Although the drugs are being manufactured by biotechnology companies for medical use only, their supplies are being diverted to black markets where they can be purchased illegally.

3. Some known side effects are facial nerve paralysis and brain edema, while chronic use could in theory cause cancer due to uncontrolled cell growth.

43. Do Sports Drinks Work?

1. Your skin loses heat by radiation, conduction, convection, and evaporation.

2. You must drink more to counteract the effects of faster sweat loss and less effective cooling.

3. During exercise and sports that last longer than an hour, 2–4 cups of these drinks diluted fifty-fifty with water should be consumed two to four hours before exercise and a cup every twenty minutes during the workout period.

44. Diet and Breast Cancer

1. According to the NRC, fat should be restricted to 30% of the diet.

2. High fiber intake may reduce the levels of circulating estrogens, which can promote the growth of breast tumors.

3. Isoflavinoids

45. Implications of Fat Replacement for Food Choice and Energy Balance

1. High relative fat intakes could specifically contribute to overeating and obesity.

2. Dietary fats are more efficiently used and more readily converted to body fat than carbohydrates.

3. Fat replacement in the diet should produce reductions in the percentage of energy from fat.

SECTION SEVEN: CONTROVERSIES AND ETHICAL ISSUES

46. Xenotransplants from Animals: Examining the Possibilities

1. Xenotransplants are animal-to-human organ transplants.

2. Baboons and pigs are animals usually favored as organ donors.

3. Two obstacles to zenotransplantation are: 1) rejection by the human body's own immune system and 2) transmission of animal diseases into the human population.

47. Is Hormone Replacement Therapy a Risk?

1. Hormone replacement therapy involves the use of estrogen alone or in combination with other hormones to control the symptoms of menopause.
2. Long-term therapy appears to be beneficial in reducing heart disease and bone loss.
3. Estrogen therapy has been linked to the development of uterine and breast cancers.

48. The Public Health Risk of Animal Organ and Tissue Transplantation into Humans

1. To lessen the rejection of their organs when transplanted into humans, animals must be transgenically modified.
2. A national system for virus screening would need to be developed because current methods may not be sensitive and specific enough for xenotransplantation work.
3. Porcine insulin, bovine thyroxin and pig heart valves are a few examples of materials derived from animals and used in humans.

49. The Mother of All Blood Cells

1. Stem cells are hematopoietic cells capable of generating an endless supply of other blood constituents like red cells, white cells, and platelets.
2. He combined a number of different purification techniques to achieve a nearly pure batch of stem cells.
3. Stem cell therapies could result in new treatments for cancer, AIDS, leukemias, and other inherited diseases.